CUSTOM SOLUTIONS MANUAL

to accompany

STATISTICS
Principles and Methods
Fourth Edition

for use in the
Department of Mathematics
Ohio Department
Athens, Ohio

WILEY
CUSTOM SERVICES

SOLUTIONS MANUAL TO ACCOMPANY

STATISTICS
Principles and Methods

Fourth Edition

RICHARD A. JOHNSON
GOURI K. BHATTACHARYYA
University of Wisconsin at Madison

JOHN WILEY & SONS, INC.
New York • Chichester • Weinheim • Brisbane • Singapore • Toronto

TABLE OF CONTENTS

Preface

PREFACE

This instructors manual is intended to provide the teacher with solutions to all of the exercises. We would appreciate receiving your comments, corrections and suggestions for improvements.

Richard A. Johnson

Gouri K. Bhattacharyya

Chapter 1

Introduction

1.1 The *population* consists of the entire set of responses from all teenagers, 13 to 17 years old, in the United States while the *sample* consists of the responses of the particular 1055 teens contacted in the telephone survey.

1.3 The *population* is presumably the collection of semen concentration values for all males, worldwide, who lived during the study period. The *sample* is the collection of semen concentrations measured from the particular 14,847 men who took part in the study. (The speaker's dramatic remark refers to all adult males alive today.)

1.5 The newspaper is suggesting that the *population* is the collection of preferences for each adult in the city while the *sample* is the collection of preferences of the particular persons who sent in their votes. This sample is apt to be non-representative because those persons in the sample are self-selected. Only the few who feel very strongly positive will likely send in a vote.

1.7 (a) The yes/no answer regarding a monthly plan, for each of the 30 students, is the sample on which the statement is based.

    (b) This is anecdotal. No data given.

    (c) The yes/no answer regarding brand loyalty for each of the 35 customers who reported purchases is the sample on which the statement is based.

1.9 The term "too long" is not well defined. By asking a number of people, we may determine that 5 minutes is too long. Further, the time will not be the same for all people. One improved statement of purpose is:

Purpose: Determine if over half the persons take over 5 minutes to get cash during the lunch hour.

1.11 First number the classrooms 1 to 35. In Table 1, we started in row 20 using columns 29 and 30. Reading downward, and ignoring 00 and numbers above 35, we selected rooms 8, 7, 1 and 19.

1.13 We started in row 10 and read down column 9 and then down column 6 from the top. We ignored the second digit in a pair, and kept reading, where the two digits were the same. That type of assignment of students is not allowed.

20 pairs of random digits

4, 0   1, 2   4, 2   5, 2   2, 1

5, 1   3, 2   2, 0   7, 6   5, 4

0, 1   2, 5   2, 7   3, 6   5, 4

2, 4   5, 7   3, 5   6, 7   7, 2

(a) $4/20 = .20$

(b) $9/20 = .45$

(c) $7/20 = .35$

1.15

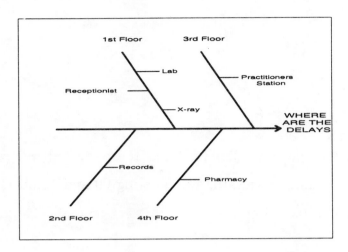

1.17 (a) The miniture poodles could never be observed even if the greatly outnumber
         the Great Danes. Only the big dogs can volunteer to show they were inside
         the fence.

      (b) Persons who call-in their opinions are self selected because they have strong
         opinions. This is analogous to the big dogs who are the only volunteers to
         show they were inside the fence.

Chapter 2

Organization and Description of Data

2.1   (a)   The percentage in other classes is $100 - 38.6 - 12.8 - 10.1 - 9.9 - 7.7 = 20.9$ %

      (b)

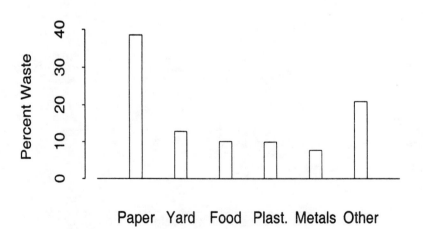

      (c)   The percentage of waste that is paper or paperboard = 38.6 %

           The percentage of waste in the top two categories is $38.6 + 12.8 = 51.4$

           The percentage in the top five categories is $38.6 + 12.8 + 10.1 + 9.9 + 7.7$

$= 79.1$

2.3 The frequency table for blood type is

Blood type	Frequency	Relative Frequency
O	16	$0.40 = 16/40$
A	18	$0.45 = 18/40$
B	4	$0.10 = 4/40$
AB	2	$0.05 = 2/40$
Total	40	1.00

2.5 (a) The table of relative frequencies for workers in the department is

Mode of Transportation	Frequency	Relative frequency
Drive alone	25	$0.625 = 25/40$
Car pool	3	$0.075 = 3/40$
Ride bus	7	$0.175 = 7/40$
Other	5	$0.125 = 5/40$
Total	40	1.000

(b) The pie chart for workers in the department is

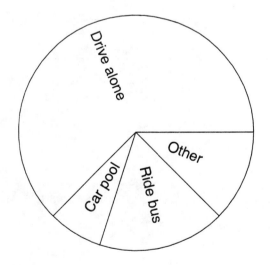

2.7 There are overlapping classes in the grouping. A report of 9 stolen bicycles will fall in two classes.

2.9 There is a gap. The response 5 close friends does not fall in any class. The last class should be 5 or more.

2.11 (a) Yes. (b) Yes. (c) Yes. (d) No. (e) No.

2.13 (a) The relative frequencies are $9/50 = .18, .48, .26$, and $.8$ for 0, 1, 2, and 3 bags , respectively.

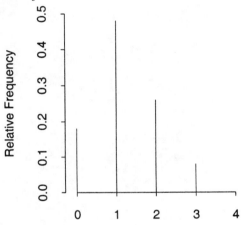

Number of bags checked

(b) Almost half of the passengers check exactly one bag. The longest tail is to the right.

(c) The proportion of passengers who fail to check a bag is $9/50 = .18$.

2.15 The dot diagram is

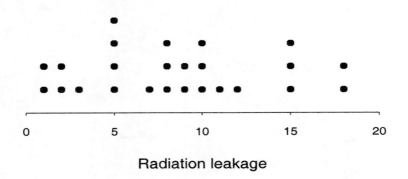

2.17 (a) The dot diagram is

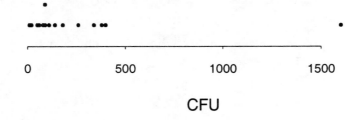

(b) There is a long tail to the right with one extremely large value 1600 CFU units

(c) There is one day so the proportion is $1/15 = .067$

2.19 (a) In the following frequency distribution of lizard speed (in meters per second), the left endpoint is included in the class interval but not the right endpoint.

Class Interval			Frequency	Relative Frequency
.45	to	.90	2	0.067
.90	to	1.35	6	0.200
1.35	to	1.80	11	0.367
1.80	to	2.25	5	0.167
2.25	to	2.70	6	0.200
		Total	30	1.001

(b) All of the class intervals are of length .45 so we can graph rectangles whose heights are the relative frequency. The histogram is

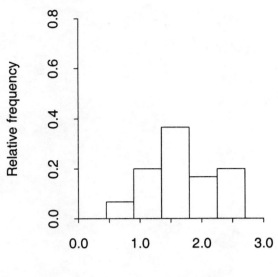

2.21 The frequencies are 5,11,13,7,8, and 5, respectively. The class intervals have unequal widths so the rectangles have

$$\text{height} = \frac{\text{relative frequency}}{\text{width of interval}}$$

For the interval 5.2 to 5.6, height = (5/49)/.4 = .0255.

For the interval 7.2 to 8.4, height = (5/49)/ 1.2 = .0850.

The histogram has a long right-hand tail.

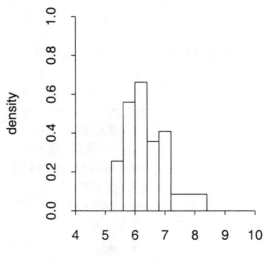

Size of Earthquake

2.23 The stem-and-leaf display of the amount of iron present in the oil is

```
0 | 6
1 | 2234455567777889
2 | 000000222445567799
3 | 022444566
4 | 1167
5 | 12
```

2.25 The double-stem display of the amount of iron present in the oil is

```
0 | 6
1 | 22344
1 | 55567777889
2 | 00000022244
2 | 5567799
3 | 022444
3 | 566
4 | 11
4 | 67
5 | 12
```

2.27 The five-stem display of the Consumer Price Index for twenty seven large cities

is

```
14 │ 55
14 │
14 │
15 │ 1
15 │ 23
15 │ 5
15 │ 66677
15 │ 889
16 │ 00
16 │ 223
16 │ 44
16 │ 7
16 │ 8
17 │ 1
17 │ 2
```

2.29 (a) The median is 3. The sample mean is

$$\bar{x} = \frac{2 + 5 + 1 + 4 + 3}{5} = \frac{15}{5} = 3$$

(b) The mean is

$$\bar{x} = \frac{26 + 30 + 38 + 32 + 26 + 31}{6} = \frac{183}{6} = 30.5$$

The ordered measurements are: 26, 26, 30, 31, 32, 38

$$\text{median} = \frac{30 + 31}{2} = 30.5$$

(c) The sample mean is

$$\bar{x} = \frac{-1 + 2 + 0 + 1 + 4 - 1 + 2}{7} = 1$$

The ordered measurements are: $-1, -1, 0, 1, 2, 2, 4$.

The median is 1.

2.31 (a) $\bar{x} = 3810/15 = 254$.

(b) The ordered observations are

10   20   50   60   80   90   90   110

140   180   260   340   380   400   1600

so the median is 110 CFU units. The one very large observation makes the sample mean much larger.

2.33 The mean is $956/12 = 79.67$. The claim ignores variability and is not true. It is certainly unpleasant with daily maximum temperature $105^0 F$ in July.

2.35 (a) $\bar{x} = 212/25 = 8.48$

(b) The sample median is 8. Since the sample mean and median are about the same, either of them can be used as an indication of radiation leakage.

2.37 The sample mean monthly rent is $4455/7 = 636.43$ dollars. The sample median is 640 dollars.

2.39 (a) $\bar{x} = 271/40 = 6.775$.

   (b) Sample median $= (6 + 7)/2 = 6.5$. Both the sample mean and the sample median give a good indication of the amount of mineral lost.

2.41 Sample median $= (176 + 187)/2 = 181.5$(minutes).

2.43 (a) The dot diagram for the diameters (in feet) of the Indian mounds in southern Wisconsin is

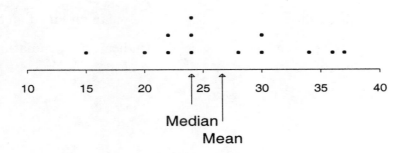

   (b) $\bar{x} = 346/13 = 26.62$. Sample median $= 24$.

   (c) $13/4 = 3.25$, so we count in 4 observations. $Q_1 = 22$ and $Q_3 = 30$.

2.45 (a) Median $= (152 + 154)/2 = 153$.

(b) $40/4 = 10$, so we need to count in 10 observations. The 11-th smallest observation also satisfies the definition.

$$Q_1 = \frac{135 + 136}{2} = 135.5, \quad Q_3 = \frac{166 + 167}{2} = 166.5$$

2.47 The ordered data are

$$
\begin{array}{ccccccccccc}
50 & 57 & 68 & 69 & 72 & 73 & 73 & 80 & 82 & 91 \\
92 & 93 & 94 & 96 & 96 & 100 & 102 & 104 & 105 & 106 \\
108 & 109 & 118 & 118 & 127
\end{array}
$$

Since the number of observations is 25, the median or second quartile is the 13th ordered observation in the list. The first quartile is the 7th observation.

$$Q_1 = 73 \quad Q_2 = 94 \quad Q_3 = 105$$

2.49   (a) The ordered observations are

$$10 \quad 20 \quad 50 \quad 60 \quad 80 \quad 90 \quad\quad 90 \quad 110$$

$$140 \quad 180 \quad 260 \quad 340 \quad 380 \quad 400 \quad 1600$$

Since the sample size is 15, the median is the 8th observation 110. To obtain Q_1, we find $15/4 = 3.75$ so the first quantile if the 4th observation in the ordered list.

$$Q_1 = 60 \quad Q_3 = 340$$

(b) The 90-th percentile requires us to count in at least $.9(15) = 13.5$ or 14 observations. The 90-th sample percentile= 400.

2.51 (a) The ordered data are 73, 74, 76, 76, 80. The median is 76^0F and the mean is $\bar{x} = 379/5 = 75.8^0F$.

(b) The mean of $(^0F - 32)$ is $\bar{x} - 32$ by property (i) of Exercise 2.50 with $c = -32$. By property (ii)

$$\text{mean of } \frac{5}{9}(^0F - 32) = \frac{5}{9}(\text{ mean of }(^0F - 32))$$

$$= \frac{5}{9}(\bar{x} - 32) = \frac{5}{9}(75.8 - 32) = 24.33^0C$$

By similar properties for the median

$$\text{median of } \frac{5}{9}(^0F - 32) = \frac{5}{9}(\text{ median of }(^0F) - 32) = \frac{5}{9}(76 - 32) = 24.44^0C$$

2.53   (a)

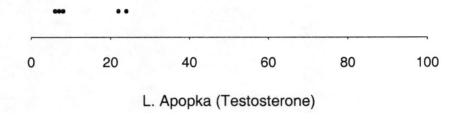

L. Apopka (Testosterone)

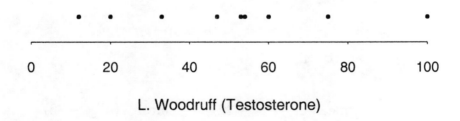

L. Woodruff (Testosterone)

(b)

$$\text{Lake Apopka } \bar{x} = \frac{67}{5} = 13.40 \quad \text{Lake Woodruff } \bar{x} = \frac{454}{9} = 50.44$$

(c) From the dot diagrams, the males in Lake Apopka have lower levels of testosterone and their sample mean is only about one-third of that for males in (un-contaminated) Lake Woodruff.

This finding is consistent with the environmentalists' concern that the contamination has affected the testosterone levels and reproductive abilities.

2.55 (a) We carry out all the necessary calculations in the following table. The mean is $\bar{x} = 12/3 = 4$.

x	$x - \bar{x}$	$(x - \bar{x})^2$
7	3	9
2	−2	4
3	−1	1
Total 12	0.0	14

(b) The variance and the standard deviation are

$$s^2 = \frac{14}{3 - 1} = 7 \quad , \quad s = \sqrt{7} = 2.646$$

2.57 (a) We carry out all the necessary calculations in the following table. The mean is $\bar{x} = 36/4 = 9$.

x	$x - \bar{x}$	$(x - \bar{x})^2$
9	0	0
7	−2	4
15	6	36
5	−4	16
Total 36	0.0	56

(b) The variance and the standard deviation are

$$s^2 = \frac{56}{4-1} = 18.67 \quad , \quad s = \sqrt{18.67} = 4.32$$

2.59 (a) We carry out all the necessary calculations in the following table.

	x	x^2
	7	49
	2	4
	3	9
Total	12	62

(b) The variance is

$$s^2 = \frac{1}{n-1}\left(\sum x^2 - \frac{(\sum x)^2}{n}\right) = \frac{1}{2}\left(62 - \frac{12^2}{3}\right)$$

$$= \frac{1}{2}(62 - 48) = 7$$

2.61   (a) $s^2 = (34 - 12^2/5)/4 = 1.30$.

     (b) $s^2 = (19 - (-7)^2/6)/5 = 2.167$.

     (c) $s^2 = (499 - 59^2/7)/6 = 0.286$.

2.63 $s = \sqrt{(9726 - 346^2/13)/12} = 6.5643$.

2.65   (a) $s^2 = (3140900 - 3810^2/15)/14 = 155{,}225.7$.

     (b) $s = \sqrt{155225.7} = 393.99$.

     (c) $s^2 = (580900 - 2210^2/14)/13 = 17{,}848.9$. so $s = \sqrt{17848.9} = 133.6$. The single very large value greatly inflates the standard deviation.

2.67   (a) $\bar{x} = 1862/10 = 186.2$.

(b) $s^2 = (353796 - 1862^2/10)/9 = 787.96$.

(c) $s = \sqrt{787.96} = 28.07$.

2.69 (a) Median $= 68.4$.

(b) $\bar{x} = 478.4/7 = 68.343$.

(c) $s^2 = (32730.34 - 478.4^2/7)/6 = 5.853$. Hence $s = 2.419$.

2.71 In Exercise 2.22, we determined that $Q_1 = 135.5$ and $Q_3 = 166.5$. Hence

$$\text{Interquartile range } = Q_3 - Q_1 = 166.5 - 135.5 = 31.0$$

2.73 No. Typically, the middle half of a data set is much more concentrated than the sum of the two quarters, one in each tail. As an example, for the water quality data of Exercise 2.17, the range is $1600 - 10 = 1590$ because of one extremely large observation. From the quartiles determined in Exercise 2.49, the interquartile range is $340 - 60 = 280$. The range is six times larger than the interquartile range.

2.75 (a) $\bar{x} = 6.775$ and $s = \sqrt{19.4096} = 4.406$.

(b) The proportion of the observations are given in the following table:

	$\bar{x} \pm s$	$\bar{x} \pm 2s$	$\bar{x} \pm 3s$
Interval:	(2.369, 11.181)	(−2.037, 15.587)	(−6.443, 19.993)
Proportion:	26/40 = 0.65	38/40 = 0.95	40/40 = 1.00
Guidelines:	0.68	0.95	0.997

(c) We observe a good agreement with the proportions suggested by the empirical guideline.

2.77 (a) $\bar{x} = 25.160$ and $s = \sqrt{114.790} = 10.714$.

(b) The proportion of the observations are given in the following table:

	$\bar{x} \pm s$	$\bar{x} \pm 2s$	$\bar{x} \pm 3s$
Interval:	(14.446, 35.874)	(3.732, 46.588)	(−6.982, 57.302)
Proportion:	36/50 = 0.72	47/50 = 0.94	50/50 = 1.00
Guidelines:	0.68	0.95	0.997

(c) We observe quite good agreement with the proportions suggested by the empirical guideline.

2.79 (a) (b) The boxplots for the salaries in City A and City B are given in Figure 2.2.

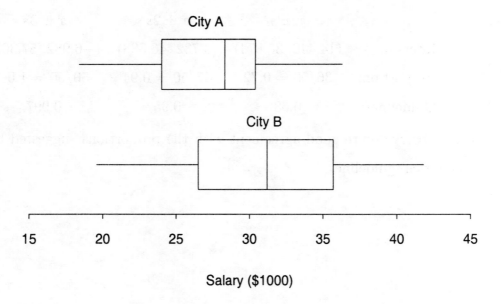

Figure 2.2: Boxplots of salaries for Exercise 2.79.

(c) The largest differences are at the upper end of the pay scale since the largest difference among the five number summaries, $41,800 - 36,300 = 5,500$ dollars, occurs at the maximum and the next largest difference, 5,300 dollars, occurs at the third quartile.

2.81 For males, the minimum and the maximum horizontal velocity of a thrown ball are 25.2 and 59.9 respectively. The quartiles are:

$Q_1 = (38.6 + 39.1)/2 = 38.85$, median $= (45.8 + 48.3)/2 = 47.05$,

$Q_3 = (49.9 + 51.7)/2 = 50.8$.

For females, the minimum and the maximum horizontal velocity of a thrown ball are 19.4 and 53.7 respectively. The quartiles are:

$Q_1 = 25.7$, median $= 30.3$, $Q_3 = 33.5$.

The boxplot of the male and female throwing speed are

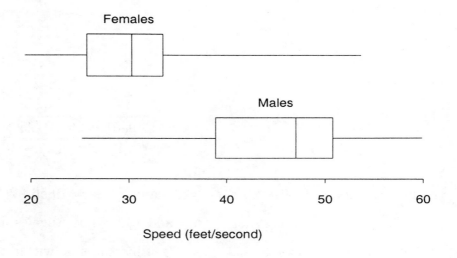

Comparing the two boxplots, we can see that males throw the ball faster than females.

2.83 (a)

$$\bar{x} = \frac{489.0}{25} = 19.56 \quad \text{and} \quad s = \sqrt{\frac{9801 - (489)^2/25}{25 - 1}} = 3.137$$

(b) Since $\bar{x} - 2s = 19.56 - 2(3.137) = 13.3$ and $\bar{x} + 2s = 19.56 + 2(3.137) = 25.8$ only the increases of 28 for Denver and 13 for Los Angeles lie outside the interval. The proportion $23/25 = .92$ of increases lie within the interval.

2.85 (a)

$$\bar{x} = \frac{341}{16} = 21.313 \quad s = \sqrt{\frac{22475 - (341)^2/16}{16 - 1}} = 31.841 \quad \text{days}$$

(b) The single large decrease, 111 days for Los Angeles, elevates the mean by almost 6 days. If Los Angeles is excluded from the calculation the sample mean reduces to $230/15 = 15.33$ days. The single large value greatly inflates the mean.

2.87 (a) The ordered data are

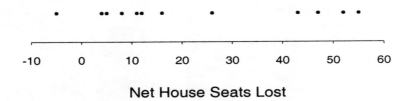

Net House Seats Lost

$$-5 \quad 4 \quad 5 \quad 8 \quad 11 \quad 12 \quad 16 \quad 26 \quad 43 \quad 47 \quad 52 \quad 55$$

median $= (12 + 16)/2 = 14$ seats lost.

(b) The maximum number of seats lost, 55, occurred when Harry S. Truman was President. The minimum number, -5 or a gain, occurred during W. Clinton's second term as President.

(c) range $= 55 - (-5) = 60$.

2.89

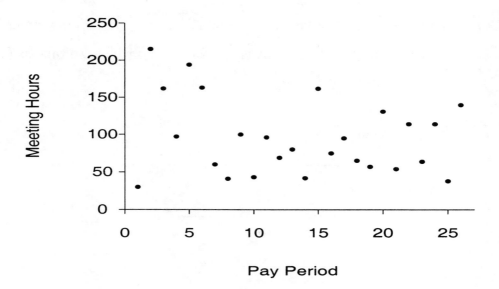

Pay Period

The value 215 from the second pay period looks high and 194 from the fifth period is possibly high.

2.91 We calculate $\bar{x} = 2501/26 = 96.2$ and $s = \sqrt{65254/25} = 51.1$ so the upper limit

is $\bar{x} + 2s = 198.4$ and the lower limit is $\bar{x} - 2s = -6.0$ which we take as 0.

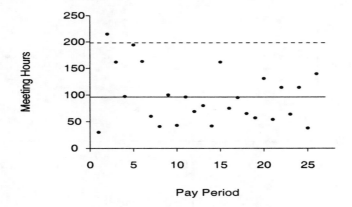

Only the value 215 from the second pay period is out of control.

2.93 We calculate $\bar{x} = 15.72/9 = 1.747$ and $s = \sqrt{.2838/8} = .188$ so the upper limit

is $\bar{x} + 2s = 2.12$ and the lower limit is $\bar{x} - 2s = 1.37$. Only the first value 2.17

is out of control but the main feature is the downward trend indicating that the exchange rate is not stable over this period of time. Control limits are not appropriate.

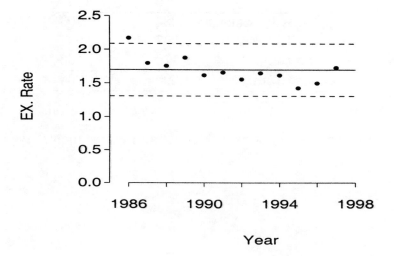

2.95 (a) The relative frequencies of the occupation groups are:

	Relative Frequency 1980	1997
Goods Producing	0.284	0.203
Service (Private)	0.537	0.637
Government	0.179	0.160
Total	1.000	1.000

(b) The proportion of persons in private service occupations has increased while the proportions in goods producing and government have decreased from 1980 to 1997.

2.97 The dot diagrams of heights for the male and female students are

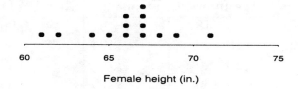

Female height (in.)

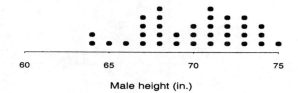

Male height (in.)

2.99 (a) The relative frequencies for the age of groom at first marriage are:

Age Interval	Frequency	Relative Frequency
16 - 18	61	0.002
18 - 20	1124	0.041
20 - 25	11768	0.428
25 - 30	9796	0.356
30 - 35	3300	0.120
35 - 45	1244	0.045
45 - 55	123	0.004
55 - 65	58	0.002
65 - 75	17	0.001
Total	27491	0.999

(b) The rectangle over the interval 18 - 20 has height $(1124/27491)/2 = 0.0204$.

The histogram of the age of groom is

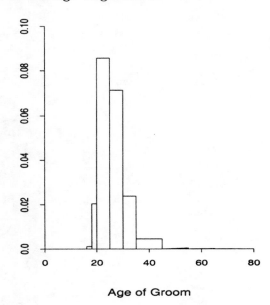

Age of Groom

(c) The proportion of grooms that marry before age 25 is

$$\frac{61 + 1124 + 11768}{27491} = .471.$$

The proportion of grooms who marry after age 30 is

$$1 - \frac{61 + 1124 + 11768 + 9796}{27491} = 1 - 0.828 = 0.172.$$

2.101 The sample mean is

$$\bar{x} = \frac{340 + 129 + 247 + 388 + 95}{5} = \frac{1199}{5} = 239.8.$$

and

$$s = \sqrt{\frac{65298.8}{5 - 1}} = 127.8 \quad \text{seconds}$$

2.103 (a) Sample median $= (9 + 9)/2 = 9$.

(b) $\bar{x} = 271/30 = 9.033$.

(c) The sample variance is

$$s^2 = \frac{1}{29}\left(2561 - \frac{271^2}{30}\right) = 3.895.$$

2.105   (a) $\bar{x} = 10$, $s = 2$.

     (b) By the properties, the new data set $x + 100$ has sample mean $= (10 + 100) = 110$ and standard deviation 2. By direct calculation, we verify

$$\frac{109 + 111 + 107 + 112 + 111}{5} = 110$$

$$\frac{(109 - 110)^2 + (111 - 110)^2 + (107 - 110)^2 + (112 - 110)^2 + (111 - 110)^2}{4}$$

$$= 4.$$

     (c) By the properties, the new data set $-3x$ has mean $= -3(10) = -30$ and standard deviation $|-3|s = 3(2) = 6$. By direct calculation, we verify

$$\frac{-27 - 33 - 21 - 36 - 33}{5} = -30$$

$$\frac{(3)^2 + (-3)^2 + (9)^2 + (-6)^2 + (-3)^2}{4} = 9\left(\frac{1 + 1 + 9 + 4 + 1}{4}\right)$$

$$= 9s^2$$

2.107 (a) The dot diagrams are

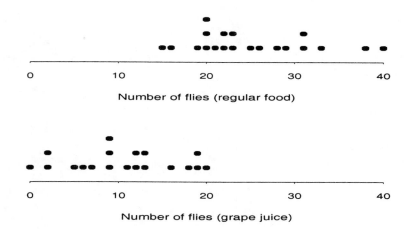

(b) From the dot diagrams we can see the number of flies (grape juice) is centered at about 11 and the number of flies (regular food) is centered near 25. The spread looks about the same.

(c) Regular food: $\bar{x} = 25.1$ and $s = 6.84$.

Grape juice: $\bar{x} = 10.6$ and $s = 6.07$.

2.109   (a) $\bar{x} = 5.38$ and $s = 3.42$.

(b) Median = 5.

(c) Range = $13 - 0 = 13$.

2.111   (a) and (b). We have $\bar{x} = 314.8/49 = 6.424$ and

$$s = \sqrt{[2047.4 - (314.8)^2/49]/48} = .721.$$

	$\bar{x} \pm s$	$\bar{x} \pm 2s$	$\bar{x} \pm 3s$
Interval:	(5.703, 7.15)	(4.98, 7.87)	(4.26, 8.59)
Proportion:	$38/49 = 0.776$	$46/49 = 0.939$	$49/49 = 1.000$
Guidelines:	0.68	0.95	0.997

(c) The count 38 is a little high. The histogram shows a long right-hand tail. The shorter left-hand tail accounts for the high count in the first interval.

2.113 (a) Median = 4.505, $Q_1 = 4.30$ and $Q_3 = 4.70$.

(b) 90-th percentile $= (4.80 + 5.07)/2 = 4.935$.

(c) $\bar{x} = 4.5074$ and $s = 0.368$.

(d) The boxplot of acid rain in Wisconsin is

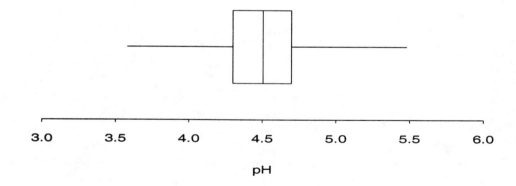

pH

2.115 (a) Median $= 6.3$, $Q_1 = 5.9$ and $Q_3 = 6.9$.

(b) $\bar{x} = 314.8/49 = 6.424$ and $s = \sqrt{[2047.4 -- (314.8)^2/49]/48} = .721$.

(c)

Class Interval (%)	Frequency	Relative Frequency
(5.2, 5.6]	5	0.1020
(5.6, 6.0]	11	0.2245
(6.0, 6.4]	13	0.2653
(6.4, 6.8]	7	0.1429
(6.8, 7.2]	8	0.1633
(7.2, 8.4]	5	0.1020
Total	49	1.0000

We use the convention that the right endpoint is included in the class interval.

(d) The boxplot of the data is

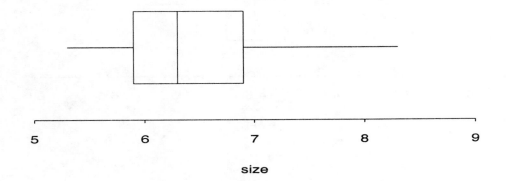

2.117 (a) The dot diagram of winning times of men's 400-meter freestyle in Olympics (1908-1992) is

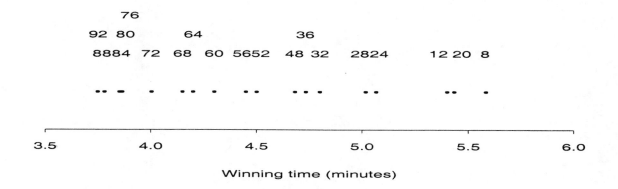

(b) It is not reasonable, because a frequency distribution would not show the systematic decrease of the winning times over the years, which is the main feature of these observations.

2.119 (a) The time series plots is given in Figure 2.3.

(b) There seems to be a systematic decrease and then increase in level.

2.121   (a) The partial MINITAB output is

Variable	N	Mean	Median	StDev
Speed	30	1.724	1.665	0.573

Variable	Minimum	Maximum	Q1	Q3
Speed	0.500	2.670	1.288	2.125

(b) The partial MINITAB output for the acid rain data.

	N	MEAN	MEDIAN	STDEV
PH	50	4.5074	4.5050	0.3681

	MIN	MAX	Q1	Q3
PH	3.5800	5.4800	4.2950	4.7000

(Note that MINITAB uses a slightly different convention for determining Q_1 and Q_3.)

2.123 The partial MINITAB output for the data set in Table 4.

Variable	N	Mean	Median	StDev
Booksal	40	306.7	301.7	143.3

Variable	Minimum	Maximum	Q1	Q3
Booksal	16.0	621.4	217.5	426.4

2.125 The mean and standard deviation given by MINITAB are the rounded off values of the answer given by SAS.

2.127   (a) The histogram of the alligator data is

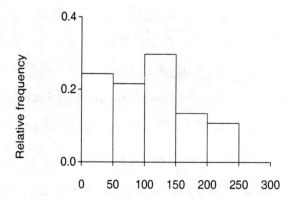

LHE2(pg/ml)

(b)

$$\bar{x} = \frac{4035}{37} = 109.1 \quad s = \sqrt{\frac{155672}{37-1}} = 65.8$$

2.129 (a) Descriptive Statistics: Maltextr

Variable	N	Mean	Median	StDev
Maltextr	40	77.458	77.400	1.101

(b) The ordered observations are

75.3 75.7 75.9 75.9 76.2 76.3 76.4 76.4 76.6 76.6

76.7 76.9 76.9 77.0 77.0 77.1 77.4 77.4 77.4 77.4

77.4 77.5 77.6 77.6 77.8 77.9 77.9 77.9 77.9 77.9

78.0 78.1 78.3 78.4 78.4 78.5 79.1 79.2 80.0 80.4

There are 40 observations so the median $= (77.4 + 77.4)/2 = 77.4$. The first quartile is the average of the $40/4 = 10$th and 11th observation in the sorted list. $Q_1 = (76.6 + 76.7)/2 = 76.65$ and $Q_3 = (77.9 + 78.0)/2 = 77.95$.

(c) The interval $\bar{x} \pm s$ or $(76.36, 78.56)$ has relative frequency $30/40 = .75$ compared to .683. The interval $\bar{x} \pm 2s$ or $(75.26, 79.66)$ has relative frequency $38/40 = .95$ compared to .95. The interval $\bar{x} \pm 3s$ or $(74.16, 80.76$ has relative frequency 1 compared to .997. The agreement is quite good.

Chapter 3

DESCRIPTIVE STUDY OF

BIVARIATE DATA

3.1 (a) The table, with completed marginal totals, is:

| | Degree of Nausea | | | | |
	None	Slight	Moderate	Severe	Total
Pill	43	36	18	3	100
Placebo	19	33	36	12	100
Total	62	69	54	15	200

(b) The relative frequencies, by row, are:

| | Degree of Nausea | | | | |
	None	Slight	Moderate	Severe	Total
Pill	.43	.36	.18	.03	1.00
Placebo	.19	.33	.36	.12	1.00

(c) A much higher proportion, .43, of pill takers avoided nausea as compared to the proportion .19 among those who took the placebo. Also the proportion of persons suffering moderate and severe nausea was much lower among those receiving the pill.

3.3 To compare aging, we find the relative frequencies, by row.

Model	≤ 20 Years	> 20 Years	Total
B7	.423	.577	1.000
B27	.736	.264	1.000
B37	.991	.009	1.000

The proportion of planes over 20 years old is highest for the (oldest) model B7 and it decreases, to a negligible amount for the (newest) model B37.

3.5 (a) The two-way frequency table is:

		Iron		Total
		Low	High	
Alkalinity	Low	8	2	10
	High	4	5	9
	Total	12	7	19

(b) The relative frequencies are:

		Iron		Total
		Low	High	
Alkalinity	Low	.421	.105	.526
	High	.211	.263	.474
	Total	.632	.368	1.000

(c) The relative frequencies, by row, are:

		Iron		Total
		Low	High	
Alkalinity	Low	.800	.200	1.000
	High	.444	.556	1.000

3.7 (a) The two-way frequency table is:

	Major				Total
	B	H	P	S	
Male	12	4	5	14	35
Female	6	0	4	4	14
Total	18	4	9	18	49

(b) The relative frequencies are:

	Major				Total
	B	H	P	S	
Male	.245	.082	.102	.286	.715
Female	.122	0	.082	.082	.286
Total	.367	.082	.184	.368	1.001

(Alternate solution using the computer)

When a data set is large, it useful to enter it once on a computer and instruct it to do the counting. To do so, we code gender and intended major as numbers. We choose 0 if male, 1 if female and

$$1 = B , 2 = H , 3 = P , 4 = S.$$

With the coded data in a file called 2.96.dat,

```
       ROWS: GENDER      COLUMNS: MAJOR

                  1          2          3          4        ALL

        0        12          4          5         14         35
        1         6          0          4          4         14
      ALL        18          4          9         18         49

      CELL CONTENTS --
                COUNT
```

We can also calculate relative frequencies by row. More precisely, 100 × (relative frequency) is obtained from the MINITAB command

TABLE C1 C4:
ROWPERCENT.

```
        ROWS: GENDER      COLUMNS: MAJOR

                    1         2         3         4       ALL
          0      34.29     11.43     14.29     40.00    100.00
          1      42.86       --      28.57     28.57    100.00
        ALL      36.73      8.16     18.37     36.73    100.00

        CELL CONTENTS --
                        % OF ROW
```

3.9 (a) The two-way frequency table with marginal totals is:

	Sales Less than 80 B	Sales 80 B or More	Total
Japan	10	6	16
U.S.	10	6	16
Total	20	12	32

(b)

	Sales Less than 80 B	Sales 80 B or More	Total
Japan	.3125	.1875	.5000
U.S.	.3125	.1875	.5000
Total	.6250	.3750	1.000

(c) From the row total proportions, we see that each of Japan and the U.S. are equally represented in the top 50 corporations. According to the column entries, both countries have the same proportion of corporations with sales above and below 80 billion dollars.

3.11 (a) The proportions, by row, for each condition are

Good Condition

	Died	Survived	Total
Research Hospital	.021	.979	1.000
Community Hospital	.027	.973	1.000

Bad Condition

	Died	Survived	Total
Research Hospital	.050	.950	1.000
Community Hospital	.070	.930	1.000

(b) The research hospital has a higher proportion of patients in good condition that survive, .979 versus .973, and a higher proportion of patients in poor condition that survive, .950 versus .930. Whether you were in bad or in good condition, you should prefer the research hospital.

(c) We have reached just the opposite conclusion of that reached in Exercise 3.10. In this example of Simpson's paradox, the condition of the patient acted as the lurking variable. The proportion of patients in poor condition is much higher at the research hospital so that kept down their overall survival rate calculated in Exercise 3.10.

3.13 (a) Of course, the fact that 21 out of 57, or proportion .368 quit smoking, by itself, would seem to be stronger evidence. Intuitively, we tend to think incorrectly that no persons would have quit without the medicated patch.

(b) Most people respond positively when they are given attention. The placebo trials make it possible to treat all subjects alike except for the presence or absence of medication. Twenty percent, 11 out of 55, responded positively to the procedure, even without the medication. This makes the success of the medicated patch less spectacular but provides a proper frame of reference.

3.15 No. The value of r can be small even if there is a strong relationship along a curve as illustrated in Figure 2 of the text.

3.17 (a) No. There is a tendency for high divorce rates to correspond to low death rates but correlation does not establish causal relation.

(b) Age could be a lurking variable. States with a large proportion of the population in their twenties, thirties and forties could have high divorce rates and low death rates while states with a high proportion of retired persons may have a high death rates but low divorce rates.

3.19 (a) A computer calculation gives $r = -.460$ for males.

The scatter diagram for males and the multiple scatter plot are

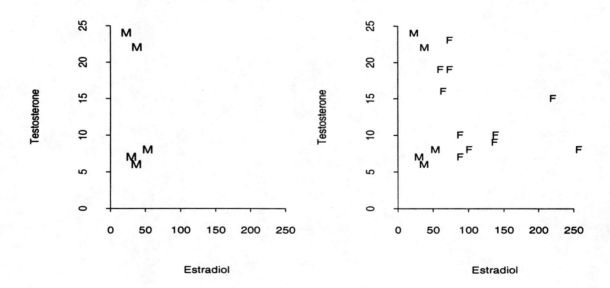

(b) A computer calculation gives $r = -.415$ for females.

(c) Both have about the same testosterone level but Females have higher levels of estradiol and are more variable.

3.21 Only Figure 6(c) has a northwest-southeast pattern indicating a negative value for r. Since the tightest pattern, indicating the largest r, is in Figure 6(a) the matches are:

(a) $r = -.3$ and Figure 6(c), (b) $r = .1$ and Figure 6(b),

(c) $r = .9$ and Figure 6(a).

3.23 Identifying the sums of squares about the means as S_{xx}, S_{yy}, and S_{xy}, respectively, we find

$$r = \frac{S_{xy}}{\sqrt{S_{xx}}\sqrt{S_{yy}}} = \frac{-204.3}{\sqrt{530.7}\sqrt{215.2}} = -.605$$

3.25 Let x = the amount of hydrogen and y = the amount of carbon. Then, with $n = 11$, we calculate

$$\sum x = 533.80, \quad \sum y = 621.00$$

$$\sum x^2 = 43,124.84, \quad \sum xy = 43,760.84, \quad \sum y^2 = 48,624.58$$

so

$$S_{xx} = \sum x^2 - \frac{(\sum x)^2}{n} = 43,124.84 - \frac{(533.80)^2}{11} = 17,220.98$$

$$S_{yy} = \sum y^2 - \frac{(\sum y)^2}{n} = 48,624.58 - \frac{(621.00)^2}{11} = 13,566.31$$

$$S_{xy} = \sum xy - \frac{(\sum x)(\sum y)}{n} = 43,760.84 - \frac{(533.80)(621.00)}{11} = 13,625.40$$

Consequently,

$$r = \frac{S_{xy}}{\sqrt{S_{xx}}\sqrt{S_{yy}}} = \frac{13,625.40}{\sqrt{17,220.98}\sqrt{13,566.31}} = .891$$

3.27 (a) The scatter diagram, in Figure 3.2, suggests (.76, 90) may be unusual.

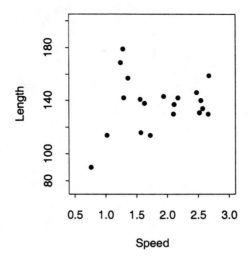

Figure 3.2: Scatter Diagram. Exercise 3.27.

(b) Let x = speed and y = body length. Then, with $n = 20$, we calculate

$$\sum x = 37.18 \quad , \quad \sum y = 2752$$

$$\sum x^2 = 75.8684 \ , \quad \sum y^2 = 386,384 \quad \sum xy = 5151.84$$

so

$$S_{xx} = \sum x^2 - \frac{\left(\sum x\right)^2}{n} = 75.8684 - \frac{(37.18)^2}{20} = 6.7508$$

$$S_{yy} = \sum y^2 - \frac{\left(\sum y\right)^2}{n} = 386,384 - \frac{(2752)^2}{20} = 7708.8$$

$$S_{xy} = \sum xy - \frac{\left(\sum x\right)\left(\sum y\right)}{n} = 5151.84 - \frac{(37.18)(2752)}{20} = 35.872$$

Consequently,

$$r = \frac{S_{xy}}{\sqrt{S_{xx}}\sqrt{S_{yy}}} = \frac{35.872}{\sqrt{7708.8}\sqrt{6.7508}} = .157.$$

3.29 Let $x =$ Boy Scout membership(millions) and $y =$ number of inmates(millions). Then, with $n = 6$, we calculate

$$\sum x = 25.74 \quad , \quad \sum y = 9.13$$

$$\sum x^2 = 110.56 \; , \quad \sum y^2 = 14.0355 \quad \sum xy = 39.2493$$

so

$$S_{xx} = \sum x^2 - \frac{(\sum x)^2}{n} = 110.56 - \frac{(25.74)^2}{6} = .1354$$

$$S_{yy} = \sum y^2 - \frac{(\sum y)^2}{n} = 14.0355 - \frac{(9.13)^2}{6} = .1427$$

$$S_{xy} = \sum xy - \frac{(\sum x)(\sum y)}{n} = 39.2493 - \frac{(25.74)(9.13)}{6} = .1266$$

Consequently,

$$r = \frac{S_{xy}}{\sqrt{S_{xx}}\sqrt{S_{yy}}} = \frac{.1266}{\sqrt{.1354}\sqrt{.1427}} = .911.$$

3.31 (a) Let x = dose and y = reaction time. Then, with $n = 8$, we calculate

$$\sum x = 63 \quad , \quad \sum y = 19.5$$

$$\sum x^2 = 665 \ , \quad \sum xy = 158.00 \ , \quad \sum y^2 = 54.79$$

so

$$S_{xx} = \sum x^2 - \frac{(\sum x)^2}{n} = 665 - \frac{(63)^2}{8} = 168.875$$

$$S_{yy} = \sum y^2 - \frac{(\sum y)^2}{n} = 54.79 - \frac{(19.5)^2}{8} = 7.259$$

$$S_{xy} = \sum xy - \frac{(\sum x)(\sum y)}{n} = 158.00 - \frac{(63)(19.5)}{8} = 4.438$$

Consequently,

$$r = \frac{S_{xy}}{\sqrt{S_{xx}}\sqrt{S_{yy}}} = \frac{4.438}{\sqrt{168.875}\sqrt{7.259}} = .127 \ .$$

(b) The scatter diagram shown in Figure 3.4 shows a strong relation but it is a curved line relationship. Here r is not appropriate because it does not capture the strength of the relationship.

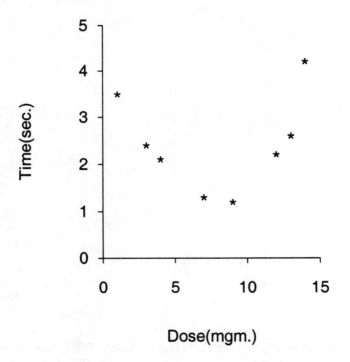

Figure 3.4: A Relationship That is Not Linear. Exercise 3.31.

3.33 (a) The scatter diagram is shown in Figure 3.6. The pattern runs from lower
           left to upper right and is not very tight. We estimate $r = .2$.

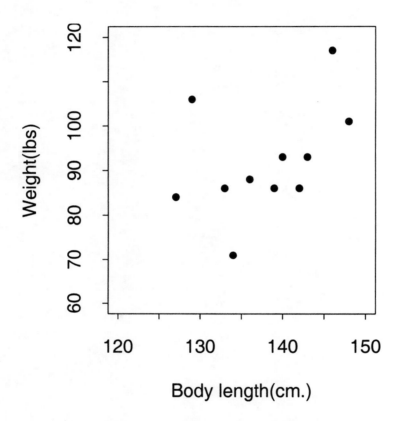

Figure 3.6: Scatter Diagram. Exercise 3.33.

(b) Let x = length and y = weight we calculate

$$\sum x = 1511 \;,\quad \sum y = 1011$$

$$\sum x^2 = 208153 \;,\quad \sum y^2 = 94453 \;,\quad \sum xy = 139141$$

so

$$S_{xx} = \sum x^2 - \frac{(\sum x)^2}{n} = 208153 - \frac{(1511)^2}{11} = 596.545$$

$$S_{yy} = \sum y^2 - \frac{(\sum y)^2}{n} = 94453 - \frac{(1011)^2}{11} = 1532.909$$

$$S_{xy} = \sum xy - \frac{(\sum x)(\sum y)}{n} = 139141 - \frac{(1511)(1011)}{11} = 266.364$$

Consequently,

$$r = \frac{S_{xy}}{\sqrt{S_{xx}}\sqrt{S_{yy}}} = \frac{266.364}{\sqrt{596.545}\sqrt{1532.909}} = .279.$$

(c) The multiple scatter diagram reveals different patterns and one possible F outlier.

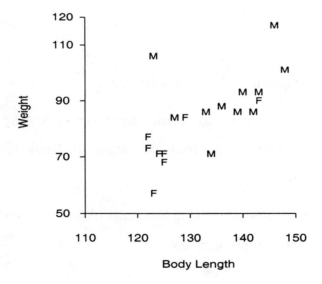

3.35 (a) The scatter diagram is shown in Figure 3.7

(b) There is a tight southwest to northeast pattern indicating a strong positive correlation. The later years have the largest garbage values and the early years the smallest.

(c) Population size is also increasing and most likely, even with more recycling, more people mean more garbage.

3.37 (a) Let $x = ($ year $- 1960)$ and $y =$ amount of garbage(mil.tons). Then, with $n = 5$, we calculate

$$\sum x = 97 \quad , \quad \sum y = 783$$

$$\sum x^2 = 2769 \ , \quad \sum y^2 = 134,603 \quad \sum xy = 18429$$

so

$$S_{xx} = \sum x^2 - \frac{(\sum x)^2}{n} = 2769 - \frac{(97)^2}{5} = 887.2$$

$$S_{yy} = \sum y^2 - \frac{(\sum y)^2}{n} = 134,603 - \frac{(783)^2}{h} = 11985.2$$

$$S_{xy} = \sum xy - \frac{(\sum x)(\sum y)}{n} = 18429 - \frac{(97)(783)}{5} = 3238.8$$

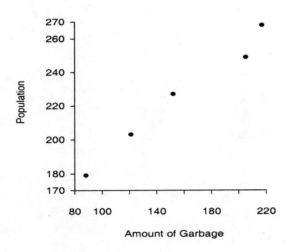

Figure 3.8: Scatter Diagram. Exercise 3.36.

Consequently,

$$r = \frac{S_{xy}}{\sqrt{S_{xx}}\sqrt{S_{yy}}} = \frac{3238.8}{\sqrt{887.2}\sqrt{11985.2}} = .993.$$

(b) The correlation is still .993. Since year is a linear transformation of (year
     − 1960)

$$\text{year} \;=\; 1 \cdot (\text{ year } - 1960) \;+\; 1960$$

The deviation for each year, or (year − $\overline{\text{year}}$) equals the same deviation
for (year − 1960) as you may verify. Consequently the sum of squares for
years and the sum of cross-products remain the same and the correlation
is unchanged.(see Exercise 3.34)

3.39 The value of y at $x = 1$ is $2 + 3(1) = 5$ and the value at $x = 4$ is $2 + 3(4) = 14$.

The line is shown in Figure 3.9. The intercept is 2, the value of y at $x = 0$, and the slope is 3, the coefficient of x.

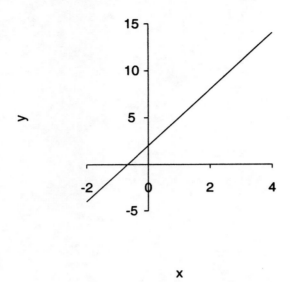

Figure 3.9: The line $y = 2 + 3x$. Exercise 3.39.

3.41 (a) $y = 10(41) - 155 = 255$.

(b) Note that $y = 0$ if $10x = 155$ or if $x = 15.5$. A profit will be made if 16 or more units are sold.

3.43 (a) (b) The scatter diagram and the visually drawn dotted line are shown in Figure 3.11(a) and (b).

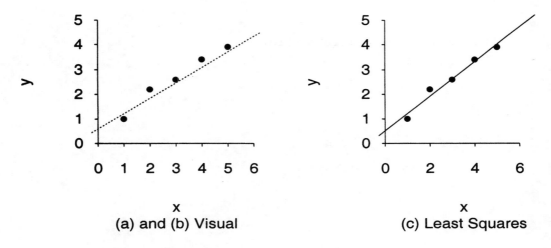

Figure 3.11: Scatter Diagram and Lines. Exercise 3.43.

(c) We use the alternative form of calculation

x	y	xy	x^2
1	1.0	1.0	1
2	2.2	4.4	4
3	2.6	7.8	9
4	3.4	13.6	16
5	3.9	19.5	25
15	13.1	46.3	55
$\bar{x} = 3$	$\bar{y} = 2.62$	$\sum xy$	$\sum x^2$

so

$$S_{xy} = \sum xy - \frac{(\sum x)(\sum y)}{n} = 46.3 - \frac{(15)(13.1)}{5} = 7.00$$

$$S_{xx} = \sum x^2 - \frac{(\sum x)^2}{n} = 55 - \frac{(15)^2}{5} = 10.00$$

and

$$\hat{\beta}_1 = \frac{S_{xy}}{S_{xx}} = \frac{7}{10} = .70$$

$$\hat{\beta}_0 = \bar{y} - \hat{\beta}_1\bar{x} = 2.62 - (.7)3 = .52$$

and the least squares line is

$$\hat{y} = .52 + .70x$$

This is the solid line in Figure 3.11(c).

3.45 (a)

$$\hat{\beta}_1 = \frac{S_{xy}}{S_{xx}} = \frac{1024}{941} = 1.088$$

$$\hat{\beta}_0 = \bar{y} - \hat{\beta}_1\bar{x} = \frac{399}{9} - (1.088)\frac{1321}{9} = -115.4$$

(b)

$$\hat{y} = -115.4 + 1.088(170) = 69.6 \text{ degrees}$$

(c)

$$r = \frac{S_{xy}}{\sqrt{S_{xx}}\sqrt{S_{yy}}} = \frac{1024}{\sqrt{941}\sqrt{1783}} = .791$$

3.47 (a) The scatter diagram is

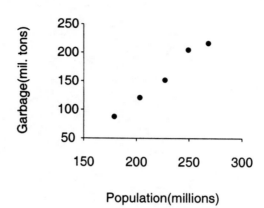

(b) We calculate $n = 5, \sum x = 1126, \sum y = 783, S_{xx} = 5,028.8$ and $S_{xy} = 7,688.4$.

$$\hat{\beta}_1 = \frac{S_{xy}}{S_{xx}} = \frac{7,688.4}{5,028.8} = 1.529$$

$$\hat{\beta}_0 = \bar{y} - \hat{\beta}_1 \bar{x} = \frac{783}{5} - (1.529)\frac{1126}{5} = -187.7$$

so the least squares line is $\hat{y} = -188 + 1.53x$.

(c) According to the straight line fit, each million persons contributes 1.53 million tons which translates into 1.53 tons per person?

3.49 The relative frequencies, by row, are:

Type of	Amount of Aid			
Representation	Increased	Unchanged	Decreased	Total
Self	.321	.587	.092	1.000
Attorney	.515	.463	.022	1.000

Having an attorney improves the chances of getting an increase.

3.51   (a) $\bar{x} = 542/30 = 18.067$

     (b) The relative frequencies, by manufacturer, are:

Manufacturer	Carbohydrates		
	Above mean	Below mean	Total
General Mills	4	6	10
Kellogg	4	6	10
Quaker	4	6	10
Total	12	18	30

(c)

Manufacturer	Carbohydrates		
	Above mean	Below mean	Total
General Mills	.4	.6	1.0
Kellogg	.4	.6	1.0
Quaker	.4	.6	1.0

The row proportions are exactly the same for each row.

3.53   (a) The frequency table is

| | Drive | | |
Size	2-Wheel	4-Wheel	Total
Small	12	21	33
Full	27	15	42
Total	39	36	75

(b) The relative frequencies are:

Size	Drive		Total
	2-Wheel	4-Wheel	
Small	.16	.28	.44
Full	.36	.20	.56
Total	.52	.48	1.00

(c) The relative frequencies, by row, are:

Size	Drive		Total
	2-Wheel	4-Wheel	
Small	.364	.636	1.000
Full	.643	.357	1.000

(d) A larger proportion of small truck purchasers prefer 4-wheel drive.

3.55 (a) The scatter diagram is shown in Figure 3.16.

(b) We calculate

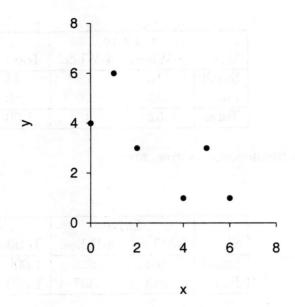

Figure 3.16: Scatter Diagram. Exercise 3.55.

x	y	$x - \bar{x}$	$y - \bar{y}$	$(x - \bar{x})(y - \bar{y})$	$(x - \bar{x})^2$	$(y - \bar{y})^2$
0	4	−3	1	−3	9	1
2	3	−1	0	0	1	0
5	3	2	0	0	4	0
4	1	1	−2	−2	1	4
1	6	−2	3	−6	4	9
6	1	3	−2	−6	9	4
18	18	0	0	−17	28	18
$\bar{x} = 3$	$\bar{y} = 3$			S_{xy}	S_{xx}	S_{yy}

so

$$r = \frac{S_{xy}}{\sqrt{S_{xx}}\sqrt{S_{yy}}} = \frac{-17}{\sqrt{28}\sqrt{18}} = -.757$$

3.57 The sum of squares are

$$S_{xy} = \sum xy - \frac{(\sum x)(\sum y)}{n} = 7,065 - \frac{(312)(289)}{20} = 2556.6$$

$$S_{xx} = \sum x^2 - \frac{(\sum x)^2}{n} = 5,848 - \frac{(312)^2}{20} = 980.8$$

$$S_{yy} = \sum y^2 - \frac{(\sum y)^2}{n} = 17,366 - \frac{(289)^2}{20} = 13190$$

so

$$r = \frac{S_{xy}}{\sqrt{S_{xx}}\sqrt{S_{yy}}} = \frac{2556.6}{\sqrt{980.8}\sqrt{13190}} = .71$$

3.59 (a) The scatter diagram is given in Figure 3.18.

(b) With $n = 6$, we calculate

$$\sum x = 82.3 \ , \quad \sum y = 56.9$$

$$\sum x^2 = 1,136.83 \ , \quad \sum xy = 785.21 \ , \quad \sum y^2 = 542.49$$

so

$$S_{xx} = \sum x^2 - \frac{(\sum x)^2}{n} = 1,136.83 - \frac{(82.3)^2}{6} = 7.948$$

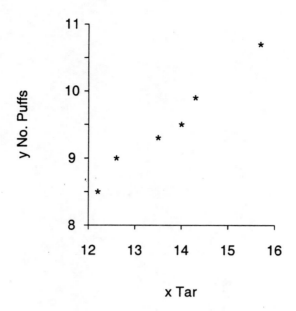

Figure 3.18: Scatter Diagram. Exercise 3.59.

$$S_{yy} = \sum y^2 - \frac{(\sum y)^2}{n} = 542.49 - \frac{(56.9)^2}{6} = 2.888$$

$$S_{xy} = \sum xy - \frac{(\sum x)(\sum y)}{n} = 785.21 - \frac{(82.3)(56.9)}{6} = 4.732$$

Consequently,

$$r = \frac{S_{xy}}{\sqrt{S_{xx}}\sqrt{S_{yy}}} = \frac{4.732}{\sqrt{7.948}\sqrt{2.888}} = .988.$$

3.61 (a) Negative; typically, the more time spent on the computer the fewer hours available for friends and other activities.

(b) Somewhat negative; Most students cram for finals and the more exams the more late night studying during finals and the fewer hours of sleep.

(c) No relation.

(d) Positive ; higher temperature tends to make people more thirsty.

3.63 (a) The scatter plot is shown in Figure 3.20.

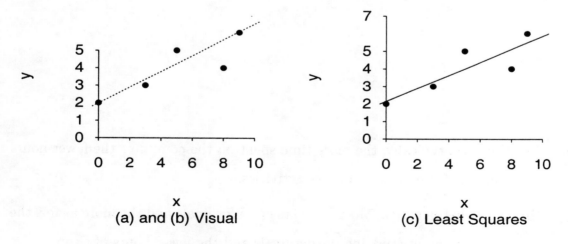

(a) and (b) Visual   (c) Least Squares

Figure 3.20: Scatter Diagram and Lines. Exercise 3.63.

(b) The visually drawn line is the dotted line in Figure 3.20(a).

(c) We calculate

x	y	$x - \bar{x}$	$y - \bar{y}$	$(x - \bar{x})(y - \bar{y})$	$(x - \bar{x})^2$
0	2	-5	-2	10	25
3	3	-2	-1	2	4
5	5	0	1	0	0
8	4	3	0	0	9
9	6	4	2	8	16
25	20	0	0	20	54
$\bar{x} = 5$	$\bar{y} = 4$			S_{xy}	S_{xx}

so

$$\hat{\beta}_1 = \frac{S_{xy}}{S_{xx}} = \frac{20}{54} = .370$$

$$\hat{\beta}_0 = \bar{y} - \hat{\beta}_1 \bar{x} = 4 - (.370)5 = 2.15$$

and the least squares line is

$$\hat{y} = 2.15 + .37x$$

This is the solid line in Figure 3.20 (c)

3.65   (a) $x =$ road roughness and $y =$ gas consumption.

   (b) $x =$ number of wins and $y =$ total sales.

   (c) $x =$ trip distance and $y =$ number of weekends at home.

3.67 At $x = \bar{x}$

$$\hat{y} = \hat{\beta}_0 + \hat{\beta}_1 \bar{x} = (\bar{y} - \hat{\beta}_1 \bar{x}) + \hat{\beta}_1 \bar{x} = \bar{y}$$

since $\hat{\beta}_0 = (\bar{y} - \hat{\beta}_1 \bar{x})$. Therefore, $(\bar{x}, \bar{y})$ is a point on the least squares line.

3.69 (a)

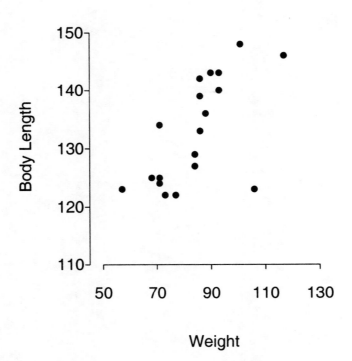

(b)

```
Correlations: bodyleng, weight


Pearson correlation of bodyleng and weight = 0.649
```

(c)

Regression Analysis: bodyleng versus weight

The regression equation is
bodyleng = 98.6 + 0.407 weight

Predictor	Coef	SE Coef	T	P
Constant	98.560	9.889	9.97	0.000
weight	0.4066	0.1157	3.51	0.003

S = 7.083

Chapter 4

PROBABILITY

4.1 (a) (ii),(v)   (e) (v)

    (b) (iv),(v)   (f) (iii),(v)

    (c) (vi)   (g) (i)

    (d) (vi)

4.3 (a) (i)   (b) (iii)   (c) (ii)

4.5   (a) $\{0, 1\}$

(b) $\{0, 1, \ldots, 344\}$

(c) $\{t : 90 < t < 425.4\}$

4.7   (a) Let us identify Bob, John, Linda, and Sue by their initials B, J, L, and S, respectively, We make a tree diagram:

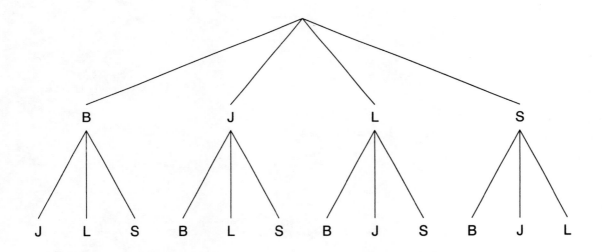

$$\mathcal{S} = \{BJ, BL, BS, JB, JL, JS, LB, LJ, LS, SB, SJ, SL\}$$

(b) A = { LB, LJ, LS },   B = { JL, LJ, JS, SJ, LS, SL }

4.9 $P(e_1) + P(e_2) + P(e_3) = .2 + .5 + .1 = .8$. Since $P(\mathcal{S}) = 1$, we must have $P(e_4) = 1 - .8 = .2$.

4.11 (a) yes

(b) no, because the sum of probabilities is less than 1

(c) yes

4.13 Denote May by e_1 and on. Because $1 + 3 + 6 + 10 = 20$, we have

$$P(e_1) = \tfrac{1}{20}, \quad P(e_2) = \tfrac{3}{20}, \quad P(e_3) = \tfrac{6}{20}, \quad P(e_4) = \tfrac{10}{20}$$

so $P(A) = P(e_1) + P(e_2) = \tfrac{4}{20} = .2.$

4.15 The relative frequencies are based on a very large number of cases and will therefore be very good approximations to the probabilities. We estimate the probability that a baby will be born in the first part of the week as

$$P(\text{ Monday }) + P(\text{ Tuesday }) + P(\text{ Wednesday }) = .146 + .153 + .148 = .447$$

4.17 (a) The tree diagram is

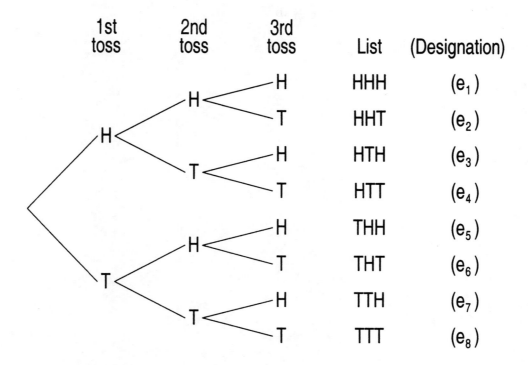

1st toss	2nd toss	3rd toss	List	(Designation)
		H	HHH	(e_1)
	H	T	HHT	(e_2)
H		H	HTH	(e_3)
	T	T	HTT	(e_4)
		H	THH	(e_5)
	H	T	THT	(e_6)
T		H	TTH	(e_7)
	T	T	TTT	(e_8)

$$\mathcal{S} = \{e_1, e_2, \ldots, e_8\}$$

(b) Assuming the coins are fair, all the elementary outcomes are equally likely.

$$P(e_1) = P(e_2) = \cdots = P(e_8) = \frac{1}{8}$$

(c) [Exactly one head] $= \{e_4, e_6, e_7\}$, its probability is $\frac{3}{8}$.

4.19 (a) Let e_1, e_2, and e_3 denote the outcomes of getting a ticket numbered 1, 2, and 3, respectively. Then $\mathcal{S} = \{e_1, e_2, e_3\}$. Since all 8 tickets are equally likely to be drawn, and there are 2 tickets with number 1, we have $P(e_1) = \frac{2}{8}$. Likewise, $P(e_2) = \frac{3}{8}$ and $P(e_3) = \frac{3}{8}$.

(b) [Odd-numbered ticket drawn] $= \{e_1, e_3\}$, so the probability is

$$P(e_1) + P(e_3) = \tfrac{2}{8} + \tfrac{3}{8} = \tfrac{5}{8} = .625.$$

4.21 (a) Each elementary outcome is a pair of numbers, the first corresponds to the white die and the second to the colored die.

$$A = \{(1,5),\ (2,4),\ (3,3),\ (4,2),\ (5,1)\}$$

$$B = \{(1,6),\ (2,5),\ (3,4),\ (4,3),\ (5,2),\ (6,1)\}$$

$$C = \{(2,6),\ (4,6),\ (6,6),\ (1,5),\ (3,5),\ (5,5),$$
$$(2,4),\ (4,4),\ (6,4),\ (1,3),\ (3,3),\ (5,3),$$
$$(2,2),\ (4,2),\ (6,2),\ (1,1),\ (3,1),\ (5,1)\}$$

$$D = \{(1,1),\ (2,2),\ (3,3),\ (4,4),\ (5,5),\ (6,6)\}.$$

(b) Probability $\tfrac{1}{36}$ for each elementary outcome.

(c) $P(A) = \tfrac{5}{36},$ $\qquad\qquad\qquad P(B) = \tfrac{6}{36} = \tfrac{1}{6}$

$\quad\ P(C) = \tfrac{18}{36} = \tfrac{1}{2},$ $\qquad\qquad P(D) = \tfrac{6}{36} = \tfrac{1}{6}$

*4.23 (a) The tree diagram is

Answer to Q_1	Answer to Q_2	List
T	T	TT
	F	TF
	I	TI
F	T	FT
	F	FF
	I	FI
I	T	IT
	F	IF
	I	II

(b) Because the student selects the answers at random, the 9 elementary out-comes in S are all equally likely, each has a probability of $\frac{1}{9}$. Let us suppose that the correct answers are T for Q_1 and T for Q_2. Then, the event "one correct answer" has the composition { TF, TI, FT, IT }, so

$$P(\text{one correct answer}) = \frac{4}{9}.$$

Note: Whatever be the correct answers for Q_1 and Q_2, there will be four cases in which one marked answer will match and one will not match.

4.25 (a) The 15 persons are equally likely to be selected. Among them there is only one of blood group AB so $P[AB] = \frac{1}{15}$.

(b) The number of person of blood group either A or B is $5 + 6 = 11$ so the required probability is $\frac{11}{15}$.

(c) $P[\text{ not } O] = \frac{5 + 6 + 1}{15} = \frac{12}{15}$.

4.27 $S = \{N, YN, YYN, YYYN, YYYYN, YYYYY\}$

4.29 (a) Letting $c, b,$ and v denote "compliance", "borderline case", and "violation", respectively,

$$\mathcal{S} = \{c_1, c_2, \ldots, c_9, b_1, b_2, b_3, v_1, v_2\}.$$

(b) The 14 elementary outcomes are equally likely, and two of them, namely v_1 and v_2, constitute the event that a violator is detected. The probability is $2/14 = .143$.

4.31 (a) The successive days cannot be considered as independent trials. The rate on one day is the same as, or is very close to, the rate on the next day. The results for successive days are not independent. There may also be a trend in rates over the year.

(b) Cars brought in with other problems are more likely to have an emission problem. For instance, there would be too many old cars in this sample.

(c) Not a representative sample. Since most air conditioners are sold during the summer, the observed relative frequency would be too small.

4.33 (a) Since the gift certificates are assigned at random, all 6 elementary outcomes are equally likely, each has probability 1/6.

(b) $P(A) = \frac{3}{6} = \frac{1}{2}$, $\qquad P(B) = \frac{2}{6} = \frac{1}{3}$.

4.35 (a) The Venn diagram is given in Figure 4.1.

(b) (i) $AB = \{e_6, e_7\}$

(ii) $\overline{B} = \{e_2, e_3, e_4, e_5\}$

(iii) $A\overline{B} = \{e_4, e_5\}$

(iv) $A \cup B = \{e_1, e_4, e_5, e_6, e_7\}$

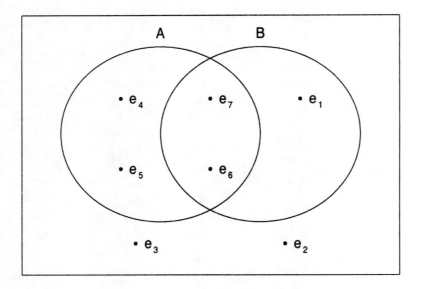

Figure 4.1: Venn diagram for Exercise 4.35(a).

4.37 (a) $\overline{C} = \{e_1, e_2, e_3, e_4, e_5, e_7\}$, $P(\overline{C}) = .72$

       (b) $AB = \{e_2, e_6, e_7\}$, $P(AB) = .12 + .14 + .14 = .40$

       (c) $A\overline{B} = \{e_1, e_5\}$, $P(A\overline{B}) = .08 + .14 = .22$

       (d) $\overline{A}\,\overline{C} = \{e_3, e_4\}$, $P(\overline{A}\,\overline{C}) = .12 + .12 = .24$

4.39 (a) Denote by e_1, e_2, e_3, and e_4 the elementary outcomes that the person hired is candidate number 1, 2, 3, and 4, respectively. The Venn diagram is given in Figure 4.2.

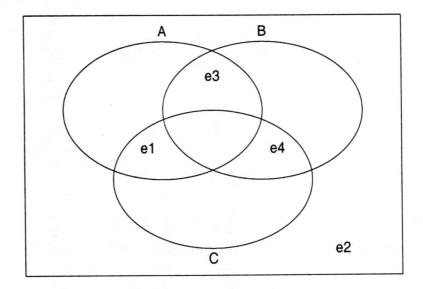

Figure 4.2: Venn Diagram for Exercise 4.39(a).

       (b) $A \cup B = \{e_1, e_3, e_4\}$ and $AB = \{e_3\}$

4.41   (a) $P(A) = .06 + .1 + .09 = .25$

$P(B) = .06 + .1 + .09 + .09 = .34$

Since $AB = \{e_5, e_8\}, \quad P(AB) = .1 + .09 = .19$

(b) $P(A \cup B) = P(A) + P(B) - P(AB) = .25 + .34 - .19 = .40$

(c) $A \cup B = \{e_1, e_5, e_8, e_2, e_9\}$

$P(A \cup B) = .06 + .1 + .09 + .06 + .09 = .40$

(d) $P(\overline{B}) = 1 - P(B) = 1 - .34 = .66$

$\overline{B} = \{e_1, e_3, e_4, e_6, e_7\}$

$P(\overline{B}) = .06 + .1 + .1 + .2 + .2 = .66.$

4.43   (a) The specified probabilities are entered in the table and underlined. The other entries are obtained in part (b).

	B	$\overline{B}$	
A	.20	.32	.52
$\overline{A}$	.16	.32	.48
	.36	.64	1.00

(b) Since $A = AB \cup A\overline{B}$, a union of mutually exclusive events, we have

$$P(A) = P(AB) + P(A\overline{B}) \quad \text{or} \quad .52 = .20 + P(A\overline{B}).$$

We get $P(A\overline{B}) = .52 - .20 = .32$. Similarly,

$$P(\overline{A}B) = .36 - .20 = .16, \qquad P(\overline{A}\,\overline{B}) = 1 - (.20 + .32 + .16) = .32$$

4.45 (a) The completed probability table is given below

	B	$\overline{B}$	
A	.20	.12	.32
$\overline{A}$	.15	.53	.68
	.35	.65	1.00

(b) $P(A\overline{B}) = .12$

(c) $P(A \cup B) = P(A) + P(B) - P(AB) = .32 + .35 - .20 = .47$

(d) $P(A\overline{B} \cup \overline{A}B) = P(A\overline{B}) + P(\overline{A}B)$, (union of incompatible events)

$$= .12 + .15 = .27$$

4.47 (a) $P(\overline{A}) = .4 + .25 = .65$

(b) $P(A\overline{B}) = .15$

(c) $P(A\overline{B}) + P(\overline{A}B) = .15 + .4 = .55$

4.49 Denoting 'violation' by V and 'compliance' by C, the classification of the 18 restaurants is shown in the following table:

		Safety		
		V	C	Total
Sanitary	V	4	3	7
	C	4	7	11
Total		8	10	18

$P(CC) = \frac{7}{18} = .389$

4.51 (a) $P(A) \quad = .08 + .02 + .20 + .10 = .40$

$P(B) \quad = .15 + .10 + .08 + .02 = .35$

$P(BC) \quad = .15 + .08 = .23$

$P(ABC) = .08.$

(b) (i) Light case and above 40.

$$P(\overline{A}\,\overline{B}) = .15 + .20 = .35$$

(ii) Either a light case or the parents are not diabetic or both.

$$P(\overline{A} \cup \overline{C}) = .15 + .10 + .15 + .20 + .02 + .10 = .72$$

(iii) A light case, age is below 40 and parents are not diabetic.

$$P(\overline{A}\,B\,\overline{C}) = .10.$$

4.53 (a) With the stated numbers identifying the gift boxes, the list is:

$$(1,1),\quad (1,2),\quad (1,3),\quad (1,4),\quad (1,5)$$
$$(2,1),\quad (2,2),\quad (2,3),\quad (2,4),\quad (2,5)$$
$$(3,1),\quad (3,2),\quad (3,3),\quad (3,4),\quad (3,5)$$
$$(4,1),\quad (4,2),\quad (4,3),\quad (4,4),\quad (4,5)$$
$$(5,1),\quad (5,2),\quad (5,3),\quad (5,4),\quad (5,5)$$

The 25 elementary outcomes are equally likely, each has the probability $\frac{1}{25}$

(b) $A = \{(2,1),\ (2,2),\ (2,3),\ (2,4),\ (2,5)$

$(3,1),\ (3,2),\ (3,3),\ (3,4),\ (3,5)$

$(1,2),\ (4,2),\ (5,2),\ (1,3),\ (4,3),\ (5,3)\},\qquad P(A) = \frac{16}{25}$

$B = \{(3,1),\ (3,2),\ (3,3),\ (3,4),\ (3,5)$

$(4,1),\ (4,2),\ (4,3),\ (4,4),\ (4,5)$

$$(5,1),\ (5,2),\ (5,3),\ (5,4),\ (5,5)$$

$$(1,3),\ (1,4),\ (1,5),\ (2,3),\ (2,4),\ (2,5)\},\qquad P(B) = \tfrac{21}{25}$$

$$AB = \{(2,3),\ (2,4),\ (2,5),\ (1,3)$$

$$(3,1),\ (3,2),\ (3,3),\ (3,4),\ (3,5)$$

$$(4,2),\ (4,3),\ (5,2),\ (5,3)\},\qquad\qquad P(AB) = \tfrac{13}{25}$$

4.55 (a) $P(B|A) = \dfrac{P(AB)}{P(A)} = \dfrac{.35}{.68} = .515$

(b) $P(\overline{B}|A) = \dfrac{P(A\overline{B})}{P(A)}$

From $P(A) = .68$ and $P(AB) = .35$ and the fact that $A = AB \cup A\overline{B}$ (union of mutually exclusive events), we calculate $P(A\overline{B}) = .68 - .35 = .33$.

Therefore,

$$P(\overline{B}|A) = \tfrac{.33}{.68} = .485$$

Alternatively, note that B and $\overline{B}$ are complementary events so

$$P(\overline{B}|A) = 1 - P(B|A) = 1 - .515 = .485$$

(c) $P(B|\overline{A}) = \dfrac{P(\overline{A}B)}{P(\overline{A})}$

Calculating $P(\overline{A}) = 1 - P(A) = 1 - .68 = .32$

$$P(\overline{A}B) = P(B) - P(AB) = .55 - .35 = .20$$

We have

$$P(B|\overline{A}) = \frac{.20}{.32} = .625.$$

4.57 $P(B|A) = \dfrac{P(AB)}{P(A)} = \dfrac{.001}{.101} = .0099$

$P(B) = .05 + .001 = .051$

Since $P(B) \neq P(B|A)$, the events A and B are not independent.

4.59　(a)　$P(\overline{A}) = 1 - P(A) = 1 - .4 = .6$

　　(b)　$P(AB) = P(B)P(A|B) = .25 \times .7 = .175$

　　(c)　$P(A \cup B) = P(A) + P(B) - P(AB) = .4 + .25 - .175 = .475.$

4.61

$$\begin{array}{|c|}\hline \text{2　Green} \\ \text{3　Red} \\ \hline \end{array} \quad \longrightarrow 2, \text{ without replacement}$$

$$5$$

(a) We denote G for green, R for red and attach subscripts to identify the order of the draws. Since the event A, a green ball appears in the first draw, has nothing to do with the second draw, we identify $A = G_1$ so

$$P(A) = P(G_1) = \frac{2}{5} = .4$$

The event $B = G_2$ is the union of G_1G_2 and R_1G_2.

$$P(G_1G_2) = P(G_1)P(G_2|G_1) = \frac{2}{5} \times \frac{1}{4} = \frac{2}{20}$$

$$P(R_1G_2) = P(R_1)P(G_2|R_1) = \frac{3}{5} \times \frac{2}{4} = \frac{6}{20}$$

Hence, $P(B) = P(G_2) = \frac{2}{20} + \frac{6}{20} = \frac{8}{20} = .4$

(b) $P(AB) = P(G_1G_2) = \frac{2}{20} = .1.$ On the other hand,

$P(A)P(B) = .4 \times .4 = .16$, and this is different from $P(AB)$.

Therefore, A and B are not independent.

4.63 We use the symbols M for male, F for female, U for unemployed and E for employed.

(a) $P(M) = .6$

$P(U|M) = .051$

$P(U|F) = .043$

(b) $P(UM) = .051 \times .6 = .0306$, and $P(UF) = .043 \times (1 - .6) = .0172$,

Adding these we obtain $P(U) = .0306 + .0172 = .0478$ so the overall rate of unemployment is 4.8%.

(c) To find $P(F|U)$, we use the results $P(U) = .0478$ and $P(UF) = .0172$ and obtain

$$P(F|U) = \frac{P(FU)}{P(U)} = \frac{.0172}{.0478} = .360.$$

4.65 (a) If independent, $P(AB) = P(A)P(B) = .6 \times .22 = .132$ so

$$P(A \cup B) = P(A) + P(B) - P(AB) = .6 + .22 - .132 = .688$$

(b) If mutually exclusive, $P(AB) = 0$ so

$$P(A \cup B) = .6 + .22 = .82$$

(c) If A and B are mutually exclusive, we have $A\overline{B} = A$ so

$$P(A|\overline{B}) = \frac{P(A\overline{B})}{P(\overline{B})} = \frac{.6}{.78} = .769.$$

4.67 (a) For simplicity, suppose there are 100 patients altogether. Then the stated percentages are precisely the numbers of patients in the designated categories.

$$\text{No. of patients below forty} = 15 + 10 + 8 + 2 = 35$$

Of these 35 patients, $8 + 2 = 10$ have a serious case. If one patient is chosen at random from these 35 patients, the probability of a serious case is $\frac{10}{35}$.

To interpret this as a conditional probability, we use the event notation B = below 40, A = serious case. The required probability is

$$P(A|B) = \frac{P(AB)}{P(B)} = \frac{(.08 + .02)}{(.15 + .10 + .08 + .02)} = \frac{.10}{.35} = \frac{10}{35}$$

(b) We have $P(B) = .35$ and

$$P(\overline{A}B) = P \text{ (below 40 and light case)} = .15 + .10 = .25$$

Consequently, $P(\overline{A}|B) = \frac{P(\overline{A}B)}{P(B)} = \frac{.25}{.35} = \frac{5}{7}$.

If a patient is chosen at random from those who are below 40, the probability that this patient will have a light case is $\frac{5}{7}$.

$P(A) = .40$ (see 4.51(a))

$P(AC) = P(\text{serious case and parents diabetic}) = .08 + .20 = .28$

$P(C|A) = \dfrac{P(AC)}{P(A)} = \dfrac{.28}{.40} = .7$

If a patient is chosen at random from those having a serious case, the probability that this patient has diabetic parents is .7.

4.69 We use the symbol D for defective and G for good, and attach subscripts to identify the order of the selection.

(a) Here the second selection is irrelevant. If one air conditioner is selected at random, the probability of its being defective is

$$P(D_1) = \tfrac{3}{15} = .2.$$

(b) $P(D_1 G_2) = P(D_1)P(G_2|D_1) = \tfrac{3}{15} \times \tfrac{12}{14} = \tfrac{36}{210} = .171.$

(c) $P(D_1 D_2) = P(D_1)P(D_2|D_1) = \tfrac{3}{15} \times \tfrac{2}{14} = \tfrac{6}{210} = .0286.$

(d) The event D_2 is the union of two incompatible events:

$$D_2 = D_1 D_2 \cup G_1 D_2 \text{ so } P(D_2) = P(D_1 D_2) + P(G_1 D_2).$$

We have already calculated $P(D_1 D_2)$ in part (c). In the same way

$P(G_1 D_2) = P(G_1)P(D_2|G_1) = \tfrac{12}{15} \times \tfrac{3}{14} = \tfrac{36}{210}.$ Therefore

$$P(D_2) = \tfrac{6}{210} + \tfrac{36}{210} = \tfrac{42}{210} = \tfrac{3}{15} = .2.$$

(e) [exactly one defective] $= D_1 G_2 \cup G_1 D_2.$

We have already found that $P(D_1 G_2) = P(G_1 D_2) = \tfrac{36}{210}.$

Hence P (exactly one defective) $= \tfrac{36}{210} + \tfrac{36}{210} = \tfrac{72}{210} = .343.$

4.71 The classification of the 20 rats is shown in the following table:

	Infected (I)	Not infected (N)	Total
Male (M)	7	5	12
Female (F)	2	6	8
Total	9	11	20

(a) There are 9 infected rats of which 2 are females. Therefore,

$$P(F|I) = \tfrac{2}{9}$$

Alternatively, we can use the definition of conditional probability

$$P(F|I) = \frac{P(FI)}{P(I)} = \frac{2/20}{9/20} = \tfrac{2}{9}$$

(b) There are 12 males of which 7 are infected, so

$$P(I|M) = \tfrac{7}{12}.$$

(c) $P(I) = \tfrac{9}{20}, \quad P(M) = \tfrac{12}{20}, \quad P(IM) = \tfrac{7}{20} = .35$

$$P(I)P(M) = \tfrac{9}{20} \times \tfrac{12}{20} = .27$$

Since $P(IM) \neq P(I)P(M)$, the events are not independent.

4.73 (a) BC, $P(BC) = 0$ since B and C are incompatible.

(b) $A \cup B$, $P(A \cup B) = P(A) + P(B) - P(AB)$

Now $P(AB) = P(A)P(B)$ by independence

$$= .7 \times .2 = .14$$

Hence $P(A \cup B) = .7 + .2 - .14 = .76$

(c) $\overline{B}$, $P(\overline{B}) = 1 - P(B) = 1 - .2 = .8$

(d) ABC, $P(ABC) = 0$ since B and C are incompatible and so are AB and C.

4.75 Denote the events

$$S \quad = \quad \text{the cooling system functions}$$
$$S_1 \quad = \quad \text{the primary unit functions}$$
$$F_1 \quad = \quad \text{the primary unit fails}$$
$$S_2(F_2) \quad = \quad \text{the back-up unit functions (fails).}$$

Then S can be expressed as the union of two incompatible events.

$$S = S_1 \cup (F_1 S_2), \qquad P(S) = P(S_1) + P(F_1 S_2)$$

We have $P(S_1) = .999$, and

$$P(F_1 S_2) = P(F_1)P(S_2|F_1) = .001 \times .910 = .00091.$$

Therefore, $P(S) = .999 + .00091 = .99991.$

4.77 (a) Denoting the success and failure in each test by S and F, respectively, the sample space is conveniently listed with the tree diagram above.

Test 1	Test 2	Test 3	List	Probability
	S	S	SSS	0.512
		F	SSF	0.128
S	F	S	SFS	0.128
		F	SFF	0.032
	S	S	FSS	0.128
F		F	FSF	0.032
	F	S	FFS	0.032
		F	FFF	0.008
				1.000

To calculate the probabilities we note that $P(S) = .8$ and $P(F) = 1 - .8 = .2$. Since the tests are independent, we calculate

$$P(SSS) = P(S)P(S)P(S) = .8 \times .8 \times .8 = .512$$

$$P(SSF) = P(S)P(S)P(F) = .8 \times .8 \times .2 = .128, \text{ etc.}$$

The results are shown in the column of probability.

(b) P (at least two successes) $= P(2 \text{ successes}) + P(3 \text{ successes})$

$$= (.128 + .128 + .128) + .512 = .896.$$

*4.79 (a) The following probabilities are specified:

$$P(D) = .03, \qquad P(N) = 1 - P(D) = 1 - .03 = .97$$

$$P(+|N) = .10, \qquad P(-|N) = 1 - P(+|N) = 1 - .10 = .90$$

$$P(-|D) = .05, \qquad P(+|D) = 1 - P(-|D) = 1 - .05 = .95$$

Using the above probabilities, we calculate

$$P(D+) = P(D)P(+|D) = .03 \times .95 = .0285$$

$$P(D-) = P(D)P(-|D) = .03 \times .05 = .0015$$

$$P(N+) = P(N)P(+|N) = .97 \times .10 = .0970$$

$$P(N-) = P(N)P(-|N) = .97 \times .90 = .8730$$

These are entered in the probability table

	+	−	Total
D	.0285	.0015	.0300
N	.0970	.8730	.9700
Total	.1255	.8745	1.0000

(b) The required probability is the conditional probability of D given $+$

$$P(D|+) = \frac{P(D+)}{P(+)} = \frac{.0285}{.1255} = .227.$$

4.81 Denote the events

$$S = \text{strep-throat} \quad \text{and} \quad A = \text{allergy}$$

We are given that $P(S) = .25, P(A) = .4,$ and $P(SA) = .1.$

(a) $P(S \cup A) = P(S) + P(A) - P(SA) = .25 + .4 - .1 = .55$

(b) $P(S)P(A) = .25 \times .4 = .10 = P(SA)$ so the events are independent.

4.83 (a) $\binom{8}{4} = \frac{8 \times 7 \times 6 \times 5}{4 \times 3 \times 2 \times 1} = 70$

(b) $\binom{10}{3} = \frac{10 \times 9 \times 8}{3 \times 2 \times 1} = 120$

(c) $\binom{20}{2} = \frac{20 \times 19}{2 \times 1} = 190$

(d) $\left(\begin{array}{c} 20 \\ 18 \end{array} \right) = \left(\begin{array}{c} 20 \\ 2 \end{array} \right) = 190$ (see part c)

(e) $\left(\begin{array}{c} 30 \\ 3 \end{array} \right) = \dfrac{30 \times 29 \times 28}{3 \times 2 \times 1} = 4060$

(f) $\left(\begin{array}{c} 30 \\ 27 \end{array} \right) = \left(\begin{array}{c} 30 \\ 3 \end{array} \right) = 4060$ (see part e)

4.85   (a) The number of possible selections of 4 persons out of 10 persons is

$$\left(\begin{array}{c} 10 \\ 4 \end{array} \right) = \frac{10 \times 9 \times 8 \times 7}{4 \times 3 \times 2 \times 1} = 210$$

(b) The number of possible selections of 2 men out of 6 men is

$$\left(\begin{array}{c} 6 \\ 2 \end{array} \right) = \frac{6 \times 5}{2 \times 1} = 15$$

and the number of possible selections of 2 women out of 4 women is

$$\left(\begin{array}{c} 4 \\ 2 \end{array} \right) = \frac{4 \times 3}{2 \times 1} = 6$$

Since the men can be selected in 15 ways, and for each selection of men, there are 6 ways the women can be selected, the number of possible selections of two men and two women is $15 \times 6 = 90$.

4.87 (a) The number of possible selections of 5 children out of 11 is $\binom{11}{5} = 462$

(b) No. of selections of 2 out of the 4 young males is $\binom{4}{2} = 6$.

No. of selections of 3 out of the 7 young females is $\binom{7}{3} = 35$.

Each of the 6 choices of the young males can accompany each of the 35 choices of the young females, so the number of selections of 2 males and 3 female students is

$$\binom{4}{2} \times \binom{7}{3} = 6 \times 35 = 210.$$

4.89 The number of possible samples of 5 jurors out of 17 is $\binom{17}{5} = 6188$.

Under random sampling, these 6188 possible samples are equally likely.

The number of samples where all 5 jurors selected are males is

$\binom{10}{5} = 252$. The jury selection shows discrimination since

$$P(\text{no female members}) = \frac{\binom{10}{5}}{\binom{17}{5}} = \frac{252}{6188} = .041.$$

4.91 The batch contains 4 defective and 16 good alternators. The number of possible samples of size 3 is $\binom{20}{3} = 1140$ and these are equally likely.

(a) $P(A) = \dfrac{\binom{16}{3}}{\binom{20}{3}} = \dfrac{560}{1140} = .491$

(b) $P(B) = \dfrac{\dbinom{4}{2} \times \dbinom{16}{1}}{\dbinom{20}{3}} = \dfrac{6 \times 16}{1140} = .084.$

4.93 No. The states with fewer seniors get more representation in this process than what the random selection would permit.

4.95 (a) The number of possible selections of 3 plots out of 9 is $\binom{9}{3} = 84$, and the 84 selections are equally likely. One row can be chosen in $\binom{3}{1} = 3$ ways, and within that row, 3 plots can be chosen in $\binom{3}{3} = 1$ way. Therefore, the number of choices such that the three plots are in the same row is

$$\binom{3}{1} \times \binom{3}{3} = 3 \times 1 = 3$$

The required probability $= \frac{3}{84} = .036$

(b) The number of ways one plot can be selected from row 1 is $\binom{3}{1} = 3$. Likewise, a plot can be selected from row 2 in 3 ways, and from row 3 in 3 ways. The number of possible selections of 3 plots, one in each row, is $3 \times 3 \times 3 = 27$ so the required probability $= \frac{27}{84} = .321$.

*4.97

	Row A				Row B	
4	bushy			6	bushy	
4	lean	$\longrightarrow 2$		3	lean	$\longrightarrow 2$
	8				9	

(a) There are 8 trees in row A from which 2 trees can be selected in $\binom{8}{2} = 28$ ways. Of these 28 equally likely selections, there are $\binom{4}{2} = 6$ selections in which both trees are bushy. Therefore,

$$P[\text{2 bushy trees selected in row A}] = \frac{6}{28}.$$

Similarly,

$$P[\text{2 lean trees selected in row B}] = \frac{\binom{3}{2}}{\binom{9}{2}} = \frac{3}{36}$$

By independence of the two selections, the required probability $= \frac{6}{28} \times \frac{3}{36} = .018$.

(b) Let A_0, A_1 and A_2 respectively denote the events of getting exactly 0, 1, or 2 bushy trees in row A, and let B_0, B_1, B_2 denote the corresponding events for row B. Then

$$[\text{Exactly 2 bushy}] = A_2 B_0 \cup A_1 B_1 \cup A_0 B_2$$

(union of mutually exclusive events)

$$P(A_2 B_0) = P(A_2)P(B_0) = \frac{\binom{4}{2}}{\binom{8}{2}} \times \frac{\binom{3}{2}}{\binom{9}{2}} = \frac{6}{28} \times \frac{3}{36} = .018$$

(see part (a))

$$P(A_1 B_1) = P(A_1)P(B_1) = \frac{\binom{4}{1}\binom{4}{1}}{\binom{8}{2}} \times \frac{\binom{6}{1}\binom{3}{1}}{\binom{9}{2}} = \frac{16}{28} \times \frac{18}{36} = .286$$

$$P(A_0 B_2) = P(A_0)P(B_2) = \frac{\binom{4}{2}}{\binom{8}{2}} \times \frac{\binom{6}{2}}{\binom{9}{2}} = \frac{6}{28} \times \frac{15}{36} = .089$$

Adding these probabilities we get $P[\text{exactly 2 bushy}] = .393$.

4.99

5	below thirty
7	over thirty

$\longrightarrow$ 4 randomly selected.

12

(a) The number of possible choices of 4 persons out 12 is $\binom{12}{4} = 495$.

(b) The number of choices of 3 persons below thirty and 1 over thirty is

$$\binom{5}{3} \times \binom{7}{1} = 10 \times 7 = 70$$

so the required probability $= \frac{70}{495} = .141$.

4.101

$$\begin{array}{|ll|}
\hline
6 & \text{yellow} \\
5 & \text{red} \\
\hline
\end{array} \longrightarrow 4$$

$$11$$

The number of possible samples of 4 bulbs out of 11 is $\binom{11}{4} = 330$, and all choices are equally likely.

(a) The number of ways 2 red and 2 yellow bulbs can be selected is

$$\binom{5}{2} \times \binom{6}{2} = 10 \times 15 = 150$$

so P [exactly 2 red] $= \frac{150}{330} = .455$.

(b) We calculate

$$P[2 \text{ red}] = \frac{150}{330} \text{ (done in part (a))}$$

$$P[3 \text{ red}] = \frac{\binom{5}{3} \times \binom{6}{1}}{330} = \frac{10 \times 6}{330} = \frac{60}{330}$$

$$P[4 \text{ red}] = \frac{\binom{5}{4}}{330} = \frac{5}{330}$$

Adding these probabilities,

$$P[\text{at least 2 red}] = \frac{150 + 60 + 5}{330} = \frac{215}{330} = .652$$

(c) $P[\text{all 4 red}] = \dfrac{\dbinom{5}{4}}{330} = \dfrac{5}{330}$

$P[\text{all 4 yellow}] = \dfrac{\dbinom{6}{4}}{330} = \dfrac{15}{330}$

Adding these probabilities, we get

$P\,[\text{all 4 of the same color}] = \dfrac{20}{330} = .061.$

4.103   (a) $S = \{1, 2, \ldots, 24\}$

(b) $S = \{p \,:\, p > 0\},$ p is tire pressure in psi.

(c) $S = \{0, 1, 2, \ldots, 50\}$

(d) $S = \{t \,:\, t > 0\},$ p is time in days.

4.105   (a) $A = \{23, 24\}$

(b) $A = \{p \,:\, 0 < p \leq 28\}$

(c) Since $.25 \times 50 = 12.5,$ $A = \{0, 1, \ldots, 12\}$

(d) $A = \{t \,:\, 0 < t < 500.5\}$

4.107 $S = \{p :\; 0 \leq p < 1\}$ where $p =$ percentage of alcohol in blood.

$A = \{p :\; .10 < p < 1\}$

$$P \text{ (ends in 4 sets)} = \frac{6}{16}.$$

4.109 The sample space is listed by means of a tree diagram which is given in Figure 4.4.

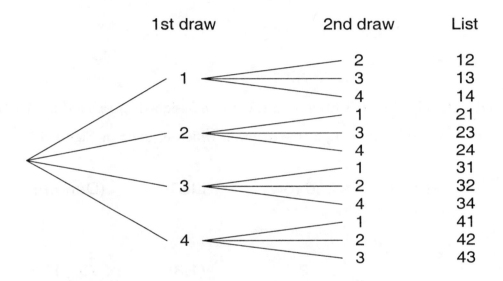

1st draw   **2nd draw**   **List**

Figure 4.4: Sample space for Exercise 4.109

(a) $\mathcal{S}$ consists of 12 elementary outcomes which are equally likely because of random selection. Of the 12 elementary outcomes, 6 correspond to even numbers so

$$P \text{ [even number]} = \frac{6}{12} = .5$$

(b) There are 9 elementary outcomes where the number is larger than 20, so the required probability is $\frac{9}{12} = .75$.

(c) There are 2 elementary outcomes, namely $\{23, 24\}$, for which the number is between 22 and 30, so the required probability is $\frac{2}{12} = .167$.

4.111 Consider the plot selected from each row to be assigned to variety 'a'. We list the sample space by drawing a tree diagram which is given in Figure 4.5.

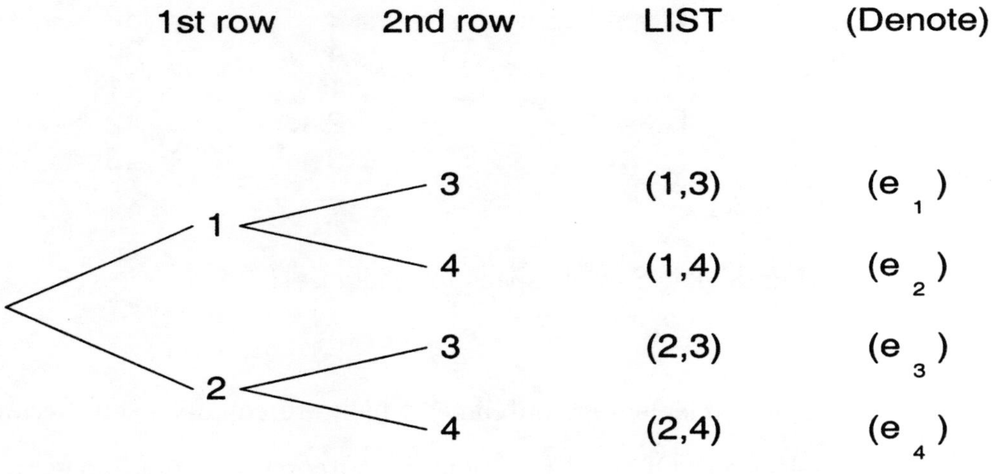
1st row
2nd row
LIST
(Denote)
1
2
3
4
3
4
(1,3)
(1,4)
(2,3)
(2,4)
(e 1)
(e 2)
(e 3)
(e 4)

Figure 4.5: Sample space for Exercise 4.111

$\mathcal{S} = \{e_1, e_2, e_3, e_4\}$ and the elements are equally likely.

[same column] $= \{(1,3), (2,4)\}$, Probability $= \frac{2}{4} = \frac{1}{2}$.

4.113 There are 12 letters which are equally likely to be selected.

(a) Since there are 5 vowels, the probability of getting a vowel is $\frac{5}{12}$.

(b) Among the 12 letters there are 3 T's so the probability of getting a T is $\frac{3}{12} = \frac{1}{4}$.

4.115 (a) $\mathcal{S} = \{125, \ 152, \ 251, \ 215, \ 512, \ 521\}$

(b) P (less than 400) $= \frac{4}{6} = \frac{2}{3}$.

(c) P (even number) $= \frac{2}{6} = \frac{1}{3}$.

4.117 (a) Either a faulty transmission or faulty brakes.

(b) Transmission, brakes and exhaust system all faulty.

(c) No faults with the transmission, brakes or the exhaust system.

(d) Either the transmission is not faulty or the brakes are not faulty.

4.119 Because $P(AB) \leq P(A) \leq P(A \cup B)$, we have $P(A) = .38, P(AB) = .2,$
$P(A \cup B) = .57.$

4.121 (a) ABC (b) $A \cup B \cup C$ (c) $AB\overline{C}$ (d) $\overline{A}\,B\overline{C}$

4.123 (a) $P(BC) = .05 + .20 = .25$ (See Figure 4.7 above.)

(b) $P(B \cup C) = P(B) + P(C) - P(BC)$ (Addition law)

$$P(B) = .05 + .10 + .20 + .15 = .50$$

$$P(C) = .05 + .05 + .20 + .18 = .48$$

$$P(BC) = .25$$

Therefore, $P(B \cup C) = .50 + .48 - .25 = .73$

Alternatively,

$$P(B \cup C) = P(B) + P(\overline{B}C) \text{ (See Figure 4.7 above.)}$$

$$= .50 + .23 = .73.$$

(c) $P(B\overline{C}) = .10 + .15 = .25$

(d) $P(A\overline{BC} \cup \overline{A}B\overline{C} \cup \overline{AB}C) = .17 + .15 + .18 = .50$

4.125 The probabilities can be determined either by using the Venn diagram or the probability table presented in the solution to Exercise 4.124.

(a) $P(B\overline{C}) = .05 + .20 = .25$

(by summing all probabilities in B but outside of C)

(b) $P(A \cup B) = P(A) + P(B) - P(AB)$ (Addition law)

$$= .51 + .45 - .17 \text{ (specified probabilities)}$$

$$= .79$$

(c) P [exactly two of the three events occur]

$= P(AB\overline{C}) + P(A\overline{B}C) + P(\overline{A}BC)$

$= .05 + .21 + .08$ (See Figure 4.8.)

$= .34.$

4.127 $P(A|B) = \dfrac{P(AB)}{P(B)} = \dfrac{.3}{.6} = .5$

$P(A)P(B) = .6 \times .5 = .30 = P(AB)$ so A and B are independent.

4.129 (a) $P(AC) = .15$

$P(A)P(C) = .6 \times .25 = .15$

Because $P(AC) = P(A)P(C)$, the events A and C are independent.

(b) $P(A\overline{B}C) = P(AC)$ (see the Venn diagram in 4.128)

$= .15$

$P(A\overline{B}) = .25 + .15 = .40$

$P(C) = .15 + .1 = .25$

$P(A\overline{B})P(C) = .40 \times .25 = .1 \neq P(A\overline{B}C)$

so the events $A\overline{B}$ and C are not independent.

4.131 (a) $P(A|\overline{B}) = \dfrac{P(A\overline{B})}{P(\overline{B})}$

From the probability table, we get

$P(A\overline{B}) = .05 + .17 = .22$

$P(\overline{B}) \;\; = .05 + .17 + .18 + .10 = .50$

so $P(A|\overline{B}) = \frac{.22}{.50} = .44$

(b) $P(B|AC) = \dfrac{P(ABC)}{P(AC)} = \dfrac{.05}{.10} = .50$

(c) From the probability table, we find $P(A) = .37, P(C) = .48, P(AC) = .10$.

Since $P(A)P(C) = .37 \times .48 = .178 \neq P(AC)$, the events A and C are

dependent.

4.133 (a) $S = \{\underbrace{11}_{e_1}, \underbrace{12}_{e_2}, \underbrace{21}_{e_3}, \underbrace{22}_{e_4}\}$

$P(e_1) = \frac{10}{15} \times \frac{9}{14} = \frac{90}{210}$ (Because the probability of a #1 marble in the

first draw is $\frac{10}{15}$, and the conditional probability of a #1 marble in the

second draw given that a #1 marble appears in the first draw, is $\frac{9}{14}$.)

Similarly, we calculate

$P(e_2) = \frac{10}{15} \times \frac{5}{14} = \frac{50}{210}$

$P(e_3) = \frac{5}{15} \times \frac{10}{14} = \frac{50}{210}$

$P(e_4) = \frac{5}{15} \times \frac{4}{14} = \frac{20}{210}$

(b) [Even number] = $\{e_2, e_4\}$. Probability = $\frac{50}{210} + \frac{20}{210} = \frac{70}{210} = \frac{1}{3}$

(c) [Larger than 15] = $\{e_3, e_4\}$, Probability = $\frac{50}{210} + \frac{20}{210} = \frac{70}{210} = \frac{1}{3}$.

4.135 (a) & (b) The tree diagram is given in Figure 4.9.

(i) $P(FF) + P(FD) + P(FN) + P(DF) + P(NF) = .64$

(ii) $P(FF) + P(FD) + P(DF) + P(DD) = .49$

4.137 Denote $G =$ Good standing, $D =$ Illegal deduction.

$$\left[\begin{array}{cc} 11 & G \\ 7 & D \end{array}\right| \longrightarrow 4$$

$$18$$

(a) The number of possible selections of 4 returns out of 18 is

$$\binom{18}{4} = 3060, \quad \text{all equally likely.}$$

$$P\left[\text{Sample has all 4 } G\text{'s}\right] = \frac{\binom{11}{4}}{\binom{18}{4}} = \frac{330}{3060} = .108.$$

(b) P [at least 2 D's] $= P[2D\text{'s}] + P[3D\text{'s}] + P[4D\text{'s}]$

$$P[2D\text{'s}] = \frac{\binom{7}{2}\binom{11}{2}}{\binom{18}{4}} = \frac{21 \times 55}{3060} = \frac{1155}{3060}$$

$$P[3D\text{'s}] = \frac{\binom{7}{3} \times \binom{11}{1}}{\binom{18}{4}} = \frac{35 \times 11}{3060} = \frac{385}{3060}$$

$$P[4D\text{'s}] = \frac{\binom{7}{4}}{\binom{18}{4}} = \frac{35}{3060}$$

Adding these we get

$$P \text{ [at least 2 } D\text{'s]} = \frac{1575}{3060} = .515.$$

4.139 (a) P (no common birthday) $= \frac{365 \times 364 \times 363}{365 \times 365 \times 365} = 1 \times \frac{364}{365} \times \frac{363}{365} = .992.$

Hence P (at least two have the same birthday) $= 1 - .992 = .008.$

(b) For each person there are 365 possible birthdays. Hence for N persons, the number of possible birthdays is

$$365 \times 365 \times \ldots \times 365(N \text{ factors}) = (365)^N.$$

In order that the birthdays of N persons to be all different,

the first person can have any of the 365 days,

the second person can have any of the remaining $365 - 1 = 364$ days,

the third person can have any of the remaining $365 - 2 = 363$ days,

. . . .

the Nth person can have any of the remaining $365 - N + 1$ days.

Hence the number of ways N persons can have all different birthdays is

$365 \times 364 \times \ldots \times (365 - N + 1)$.

Chapter 5

PROBABILITY

DISTRIBUTIONS

5.1 (a) Discrete,   (b) Continuous,   (c) Continuous,

(d) Continuous,   (e) Discrete,

5.3 (a)

Possible choices	x		Possible choices	x
{1,3}	2		{3,6}	3
{1,5}	4		{3,7}	4
{1,6}	5		{5,6}	1
{1,7}	6		{5,7}	2
{3,5}	2		{6,7}	1

(b) The distinct values of x and the corresponding probabilities are listed in the following table. All 10 choices, listed in part (a), are equally likely so

$$P[X = x] = \frac{\text{No. Choices for which } X = x}{10}$$

x	$P[X = x]$
1	.2
2	.3
3	.1
4	.2
5	.1
6	.1

5.5 (a) The ratings by the judges are given in Figure 5.1.

   (b) 4, 5, 6, 7, 8

5.7 (a) & (b) The outcomes and the values for X are given on the tree diagram in Figure 5.2.

5.9 The distinct values of X are 0, 2, 4. Since $[X = 0] = \{e_2, e_4, e_6, e_7\}$, we obtain

$$
\begin{aligned}
f(0) \;=\; P[X = 0] \;=\;& P(e_2) + P(e_4) + P(e_6) + P(e_7) \\
=\;& .29 + .09 + .12 + .13 \;=\; .63
\end{aligned}
$$

Similarly, we calculate

$$f(2) = P[X = 2] = .06 + .15 = .21$$

$$f(4) = P[X = 4] = .16$$

The probability distribution of X is given in the following table.

x	$f(x)$
0	.63
2	.21
4	.16
Total	1.00

5.11   (a) Possible values of X are 2, 3, 4, 5, 6, 7, 8, 9, 10, 11, 12.

    (b)

$$
\begin{array}{ll}
X = 2: & (1,1) \\
X = 3: & (1,2), \quad (2,1) \\
X = 4: & (1,3), \quad (2,2), \quad (3,1) \\
X = 5: & (1,4), \quad (2,3), \quad (3,2), \quad (4,1) \\
X = 6: & (1,5), \quad (2,4), \quad (3,3), \quad (4,2), \quad (5,1) \\
X = 7: & (1,6), \quad (2,5), \quad (3,4), \quad (4,3), \quad (5,2), \quad (6,1) \\
X = 8: & (2,6), \quad (3,5), \quad (4,4), \quad (5,3), \quad (6,2) \\
X = 9: & (3,6), \quad (4,5), \quad (5,4), \quad (6,3) \\
X = 10: & (4,6), \quad (5,5), \quad (6,4) \\
X = 11: & (5,6), \quad (6,5) \\
X = 12: & (6,6)
\end{array}
$$

    (c) The 36 possible outcomes are equally likely so $P[X = x]$ is the number of outcomes for which $X = x$ divided by 36. The probability distribution of X is

x	$f(x)$
2	1/36
3	2/36
4	3/36
5	4/36
6	5/36
7	6/36
8	5/36
9	4/36
10	3/36
11	2/36
12	1/36
Total	1

5.13 (a)

x	f(x)
2	1/10
3	2/10
4	3/10
5	4/10
Total	1

Yes, a probability distribution

(b)

x	f(x)
1	-1/2
2	0
3	1/2
4	1
Total	1

Not a probability distribution because $f(1)$ is negative.

(c)

x	f(x)
-2	0
-1	2/20
0	4/20
1	6/20
2	8/20
Total	1

Yes, a probability distribution.

(d)

x	f(x)
2	3/4
3	3/8
4	3/16
Total	21/16

No, because $\Sigma f(x)$ does not equal 1.

5.15 Since $P(B) = \frac{1}{2}, P(C) = \frac{1}{2}$ and purchases in different weeks are independent, the probability model is the same as that for three tosses of a fair coin. Each elementary outcome has probability $\frac{1}{8}$. Therefore

$$P[X = 0] = \frac{2}{8}, \quad P[X = 1] = \frac{4}{8}, \quad P[X = 2] = \frac{2}{8}.$$

The probability distribution of X is

x	$f(x)$
0	1/4
1	1/2
2	1/4

5.17 The selected card can have any of the numbers -3, -1, 0, 1, 3 with equal probabilities $1/5$. Define the random variable X as follows:

$X = $ the square of the selected number, can assume the values 0, 1, 9.

$[X = 0] = $ [The selected number $= 0$], $\qquad P[X = 0] = \frac{1}{5}$

$[X = 1] = $ [The selected number $= -1$ or 1], $\quad P[X = 1] = \frac{2}{5}$

$[X = 9] = $ [The selected number $= -3$ or 3], $\quad P[X = 2] = \frac{2}{5}$

The probability distribution of X is

x	$f(x)$
0	1/5
1	2/5
9	2/5
Total	1

5.19 (a) Since $3 + 1 + 1 + 1 + 1 + 3 = 10$, the probabilities of the six faces numbered 1, 2, 3, 4, 5, 6 are $\frac{3}{10}, \frac{1}{10}, \frac{1}{10}, \frac{1}{10}, \frac{1}{10}, \frac{3}{10}$, respectively. The probability distribution of X is

x	$f(x)$
1	.3
2	.1
3	.1
4	.1
5	.1
6	.3
Total	1

 (b) $P[\text{ Even number}] = f(2) + f(4) + f(6) = .1 + .1 + .3 = .5.$

5.21 Since $P[X$ is odd$] = f(1) + f(3) + f(5) = .1 + 0 + .3 = .4$ we have

$P[X$ is even$] = 1-.4 = .6$, that is, $f(2) + f(4) + f(6) = .6$.

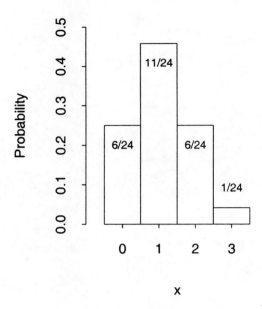

Figure 5.4: Exercise 5.20(d).

In order that $f(2), f(4)$ and $f(6)$ are all equal, and there total is .6, we must have $f(2) = .2$, $f(4) = .2$, $f(6) = .2$. The probability distribution of X is given below.

x	$f(x)$
1	.1
2	.2
3	0
4	.2
5	.3
6	.2
Total	1

5.23 (a) Consider random selection of one ball from an urn that contains the following mix of 100 numbered balls: 32 balls are numbered 2, 44 balls are numbered 4, and 24 balls are numbered 6. If X denotes the number on the selected ball, then the probability distribution of X would be as given in Table (a).

(b) Consider random selection of one ball from an urn that contains the following mix of 14 numbered balls: 3 balls are numbered −2, 4 balls are numbered 0, 5 balls are numbered 4, and 2 balls are numbered 5. If X denotes the number on the selected ball, then the probability distribution of X would be as given in Table (b).

5.25   (a) $P[X \leq 3] = f(0) + f(1) + f(2) + f(3) = .91$

      (b) $P[X \geq 2] = f(2) + f(3) + f(4) = .63$

      (c) $P[1 \leq X \leq 3] = f(1) + f(2) + f(3) = .79.$

5.27 Here $X =$ the number of customers per day.

(a) Customers will be turned away if there are 3 or more customers. The required probability is

$$P[X \geq 3] = f(3) + f(4) + f(5) = .25 + .15 + .05 = .45.$$

(b) The center's capacity is not fully utilized if fewer than 2 customers arrive. This probability is

$$P[X \leq 1] = f(0) + f(1) = .05 + .20 = .25.$$

(c) We see that

$$P[X \geq 5] = .05 \text{ and } P[X \geq 4] = .15 + .05 = .20 > .10.$$

Therefore, the capacity must be increased by 2. With a capacity of 4, the probability of turning customers away is $P[X \geq 5] = .05$.

5.29 (a) The probability histogram of X is given below.

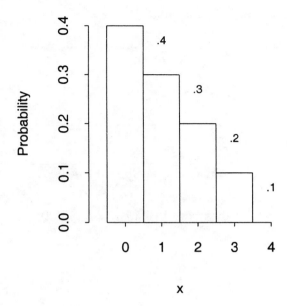

(b) The calculation is given in the following table:

x	$f(x)$	$xf(x)$	$(x - \mu)$	$(x - \mu)^2 f(x)$
1	.4	.4	-1	.4
2	.3	.6	0	0
3	.2	.6	1	.2
4	.1	.4	2	.4
Total		$2.0 = \mu$		$1.0 = \sigma^2$

$$E(X) = 2, \quad \sigma^2 = 1, \quad \sigma = 1$$

5.31 Let X denote the carpenter's net profit. Then $X = \$5,000$ with probability .2 and $X = -\$56$ (loss) with probability $1 - .2 = .8$. The probability distribution is shown in the table along with the calculation of expectation.

x	$f(x)$	$xf(x)$	
-56	.8	-44.80	
$5,000$	.2	1000.00	
Total	1	955.20	$= E(X)$

Expected return = \$955.20.

5.33

x	$f(x)$	$xf(x)$	$x^2f(x)$
0	.2401	0	0
1	.4116	.4116	.4116
2	.2646	.5292	1.0584
3	.0756	.2268	.6804
4	.0081	.0324	.1296
Total		1.2000	2.2800

$\mu = 1.2$

$\sigma^2 = 2.28 - (1.2)^2 = .84, \quad \sigma = \sqrt{.84} = .917$

5.35 (a) We denote "win" by W and "not win" by N, and attach suffixes A or B to identify the project. Listed here are the possible outcomes and calculation of the corresponding probabilities. For instance,

$$P(W_A N_B) = P(W_A)P(N_B), \text{ by independence}$$
$$= .50 \times .35 = .175$$

Outcome	Probability
$W_A W_B$	$.50 \times .65 = .325$
$W_A N_B$	$.50 \times .35 = .175$
$N_A W_B$	$.50 \times .65 = .325$
$N_A N_B$	$.50 \times .35 = .175$

(b) & (c) The amounts of profit (X) for the various outcomes are listed below.

Outcome	Profit ($) X
$W_A W_B$	$75,000 + 120,000 = 195,000$
$W_A N_B$	$75,000$
$N_A W_B$	$120,000$
$N_A N_B$	0

In the next table, we present the probability distribution of X and calculate $E(X)$.

x	$f(x)$	$xf(x)$
0	.175	0
75,000	.175	13,125
120,000	.325	3,900
195,000	.325	63,375
Total		80,400 $= E(X)$

Expected net profit $= E(X) - \text{cost} = \$80,400 - \$2,000 = \$78,400$.

5.37 (a) & (b) The expectation and standard deviation of X are calculated in the following table.

x	$f(x)$	$xf(x)$	$x^2f(x)$
0	.315	0	0
1	.289	.289	.289
2	.201	.402	.804
3	.114	.342	1.026
4	.063	.252	1.008
5	.012	.060	.300
6	.006	.036	.216
Total		1.381	3.643

$E(X)$ or $\mu = 1.381$

$\text{Var}(X)$ or $\sigma^2 = 3.643 - (1.381)^2 = 1.736$

Standard deviation of X is $\sigma = \sqrt{1.736} = 1.318$.

5.39 (a) To table the probability distribution, we calculate the function $f(x)$ for

$x = 0, 1, 2, 3$.

$$f(0) = \frac{1}{84} \begin{pmatrix} 5 \\ 0 \end{pmatrix} \begin{pmatrix} 4 \\ 3 \end{pmatrix} = \frac{4}{84}$$

$$f(1) = \frac{1}{84} \begin{pmatrix} 5 \\ 1 \end{pmatrix} \begin{pmatrix} 4 \\ 2 \end{pmatrix} = \frac{5 \times 6}{84} = \frac{30}{84}$$

$$f(2) = \frac{1}{84} \begin{pmatrix} 5 \\ 2 \end{pmatrix} \begin{pmatrix} 4 \\ 1 \end{pmatrix} = \frac{10 \times 4}{84} = \frac{40}{84}$$

$$f(3) = \frac{1}{84} \begin{pmatrix} 5 \\ 3 \end{pmatrix} \begin{pmatrix} 4 \\ 0 \end{pmatrix} = \frac{10}{84}$$

x	$f(x)$	$xf(x)$	$x^2 f(x)$
0	4/84	0	0
1	30/84	30/84	30/84
2	40/84	80/84	160/84
3	10/84	30/84	90/84
Total	1	140/84	280/84

(b) Referring to the calculations shown in the table in part (a), we find

$$\text{mean} = \Sigma x f(x) = \frac{140}{84} = 1.667$$

$$\text{variance } \sigma^2 = \Sigma x^2 f(x) - [\Sigma x f(x)]^2$$

$$= \frac{280}{84} - \left(\frac{140}{84} \right)^2 = .556$$

$$\text{standard deviation } \sigma = \sqrt{.556} = .745.$$

5.41 (a) & (b) Let us use the symbol S for a sale and N for no sale at a customer
       contact. For contact with 4 customers, we list the elementary outcomes.
       In assigning the probabilities we use the assumption of independence and
       the facts that $P(S) = .3, P(N) = .7$. For instance

$$P(SNNS) = .3 \times .7 \times .7 \times .3 = .0441$$

Elementary outcome	Probability	Value of X
$NNNN$	$(.7)^4 = .2401$	0
$NNNS$	$(.7)^3 \times .3 = .1029$	1
$NNSN$	.1029	1
$NSNN$	.1029	1
$SNNN$	.1029	1
$NNSS$	$(.7)^2 \times (.3)^2 = .0441$	2
$NSNS$	.0441	2
$SNNS$	.0441	2
$NSSN$	.0441	2
$SNSN$	.0441	2
$SSNN$	.0441	2
$NSSS$	$.7 \times (.3)^3 = .0189$	3
$SNSS$	.0189	3
$SSNS$	.0189	3
$SSSN$	.0189	3
$SSSS$	$(.3)^4 = \underline{.0081}$	4
	Total 1.0000	

$$P[X = 0] = .2401, \qquad P[X = 1] = 4 \times .1029 = .4116$$

$$P[X = 2] = 6 \times .0441 = .2646, \quad P[X = 3] = 4 \times .0189 = .0756$$

$$P[X = 4] = .0081$$

Probability distribution of X and calculation of expectation

x	$f(x)$	$xf(x)$
0	.2401	0
1	.4116	.4116
2	.2646	.5292
3	.0756	.2268
4	.0081	.0324
Total	1.0000	1.2000

(c) $E(X) = 1.2$.

5.43 $P[X \leq 2] = .4 + .3 = .7,$ and $P[X \geq 2] = .3 + .2 + .1 = .6$

Since both are $\geq .5$, median $= 2$.

5.45 (a) The Bernoulli model is not appropriate. The assumption of independence is likely to be violated because of peer pressure.

(b) The Bernoulli model is not appropriate because the measurement is on a continuous scale.

(c) The Bernoulli model is may be appropriate because two possible outcomes. Independence may be violated for items close together.

(d) Here, each house is a trial with the two possible outcomes: the delivery was on time or was not. However, the Bernoulli model does not seem plausible because of a lack of independence of the trials.

5.47 Since the event of interest is the drawing of a yellow candy, we identify S = yellow, F = any color other than yellow.

(a) Because the sampling is with replacement, and each draw has two possible outcomes S or F, the model of Bernoulli trials is appropriate. The mix has

25 candies of which 10 are yellow so the probability of drawing a yellow candy is $\frac{10}{25} = .4$. We have $p = .4$.

(b) The model of Bernoulli trials is not appropriate because the sampling is without replacement and the size of the lot is not large. The condition of independence of the trials is violated.

(c) The condition of independence of outcomes in the different trials is violated. For instance, $P(S_2|S_1) = \frac{11}{26}$ while $P(S_2|F_1) = \frac{10}{26}$. The model of Bernoulli trials is not appropriate.

5.49 Label the plots 1, 2, 3, and 4. The number of possible selections of 2 plots out of 4 is $\binom{4}{2} = 6$, and these are equally likely.

(a) Consider the event of an S in the first trial, that is, the first plot is selected. One other plot can be chosen from plots 2, 3, and 4 in $\binom{3}{1} = 3$ ways. Therefore,

$$P(S \text{ in first trial}) = \tfrac{3}{6} = \tfrac{1}{2}.$$

The same argument leads to $P(S) = \tfrac{1}{2}$ in any particular trial.

(b) Denote $S_1 S_2 = $ the event that the first and second plots are selected. We have $P(S_1 S_2) = \frac{1}{6}$. From part (a), we find that

$$P(S_1)P(S_2) = \frac{1}{2} \times \frac{1}{2} = \frac{1}{4} \neq P(S_1 S_2).$$

Therefore, the trials are not independent.

5.51 (a) Although there are two possible outcomes of each trial, the Bernoulli model is not appropriate because the 5 purchases of each consumer cannot be considered independent.

(b) Here the Bernoulli model is plausible because the 500 trials correspond to different consumers who are selected at random.

5.53 We have $P(S) = p = \frac{1}{3}, P(F) = q = 1 - \frac{1}{3} = \frac{2}{3}$.

(a) $P(FFFF) = \frac{2}{3} \times \frac{2}{3} \times \frac{2}{3} \times \frac{2}{3} = \frac{16}{81} = .1975$

(b) Because the trials are independent, the required conditional probability is the same as the (unconditional) probability of 4 trials resulting in all successes, which is

$$P(SSSS) = \left(\frac{1}{3} \right)^4 = \frac{1}{81} = .0123$$

(c) $P(FFFS) = \frac{2}{3} \times \frac{2}{3} \times \frac{2}{3} \times \frac{1}{3} = \frac{8}{81} = .0988$

5.55 (a) $P(SSSS) = .8 \times .8 \times .8 \times .8 = .4096$

(b) For a single trial $P(F) = 1 - .8 = .2$ so $P(FFFF) = (.2)^4 = .0016$

(c) Using the law of complement,

$$P \text{ [at least one success]} = 1 - P \text{ [no successes]} = 1 - .0016 \quad \text{(see (b))}$$

$$= .9984.$$

5.57 (a) The possible results in the first two trials are SS, SD, DS or DD. If SS occurs, the experiment is stopped. With SD, there is one more trial so we have either SDS or SDD. Proceeding in this way, the complete list is

$$\{SS, SDS, SDD, DSS, DSD, DDSS, DDSD, DDDS, DDDD\}$$

(b) We have $P(S) = 1/4, P(D) = 3/4$

Outcome	Probability			Value of X
SS	$(1/4)^2$	$=$	$1/16$	2
SDS	$(1/4)^2(3/4)$	$=$	$3/64$	2
SDD			$9/64$	1
DSS			$3/64$	2
DSD			$9/64$	1
$DDSS$			$9/256$	2
$DDSD$			$27/256$	1
$DDDS$			$27/256$	1
$DDDD$			$81/256$	0

(c)　　The probability distribution of X is:

x	$f(x)$
0	$81/256$
1	$126/256$
2	$49/256$
Total	1

5.59 (a) Yes. $n = 10, p = \frac{1}{6}$

(b) No, because the number of trials is not fixed.

(c) Yes. $n = 3$, and p is the probability of getting a marble numbered either 1 or 2 in a single draw so

$$p = \frac{4+3}{10} = .7.$$

(d) No, because X does not represent a count of the number of times that an event occurs.

5.61 (a) With $n = 4$, $p = .35$, $q = .65$ we obtain

$$P[X = 2] = \binom{4}{2} (.35)^2 (.65)^2 = .3105$$

(b) $n = 6$, $p = .25$, $q = .75$

$$P[X = 3] = \binom{6}{3} (.25)^3 (.75)^3 = .132$$

(c) $n = 6$, $p = .75$, $q = .25$

$$P[X = 2] = \binom{6}{2} (.75)^2 (.25)^4 = .033$$

5.63 $n = 5$, $p = .35$

$$f(x) = \binom{5}{x} (.35)^x (.65)^{5-x}$$

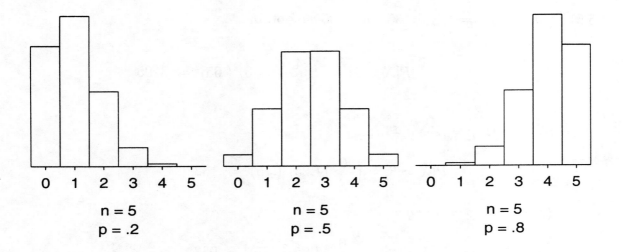

Figure 5.6: Binomial distributions for Exercise 5.62.

The calculation of $f(x)$ is presented in a table:

x	$f(x)$	
0	$(.65)^5$	$= .1160$
1	$5(.35)(.65)^4$	$= .3124$
2	$10(.35)^2(.65)^3$	$= .3364$
3	$10(.35)^3(.65)^2$	$= .1811$
4	$5(.35)^4(.65)$	$= .0488$
5	$(.35)^5$	$= .0053$
	Total	1.0000

(a) $P[X \leq 3] = f(0) + f(1) + f(2) + f(3) = .9459$

(b) $P[X \geq 3] = f(3) + f(4) + f(5) = .2352$

(c) $P[X = 2 \text{ or } 4] = f(2) + f(4) = .3852$

5.65 $n = 4$, $p = .45$, $q = .55$

$$f(x) = \binom{4}{x} (.45)^x (.55)^{4-x}$$

To calculate the required probabilities, it would be convenient to calculate all the $f(x)$ values.

x	$f(x)$	
0	$(.55)^4$	$= .0915$
1	$4(.45)(.55)^3$	$= .2995$
2	$6(.45)^2(.55)^2$	$= .3675$
3	$4(.45)^3(.55)$	$= .2005$
4	$(.45)^4$	$= .0410$
	Total	1.000

(a) $P[X \geq 3] = f(3) + f(4) = .2415$

(b) $P[X \leq 3] = 1 - f(4) = .9590$

(c) "Two or more failures" means two or fewer successes so the required probability is

$$P[X \leq 2] = f(0) + f(1) + f(2) = .7585$$

5.67 Identify S : severe leaf damage.

X = No. trees with severe leaf damage in a random sample of 5 trees.

X has a binomial distribution with $n = 5, p = .15, q = .85$.

$$f(x) = \binom{5}{x} (.15)^x (.85)^{5-x}, \quad x = 0, 1, \ldots, 5$$

(a) $P[X = 3] = f(3) = \binom{5}{3} (.15)^3 (.85)^2 = .024$

(b) $P[X \leq 2] = f(0) + f(1) + f(2) = .974$.

5.69 (a) We use the binomial table for $n = 13, p = .3$.

$$P[X = 4] = P[X \leq 4] - P[X \leq 3]$$

$$= .654 - .421 = .233$$

(b) '8 failures in 13 trials' means 5 successes in 13 trials. We use the binomial table for $n = 13$, $p = .7$

$$P[X = 5] = .018 - .004 = .014$$

(c) Using the binomial table for $n = 13$, $p = .3$, we obtain

$$P[X = 8] = .996 - .982 = .014$$

Refer to (b), and consider interchanging the names 'success' and 'failure'. Then the new p would be .3, (i.e., the old q), and the specified event would then be '8 successes in 13 trials' which is precisely the statement in part (c).

5.71 (a) Denote X = no. of successes in 5 trials.

The event 'more than 5 trials are needed in order to obtain 3 successes' means that at most two successes are obtained in 5 trials, that is, $X \leq 2$.

Since X has the binomial distribution with $n = 5, p = .7$, we use the binomial table to find the required probability

$$P[X \leq 2] = .163.$$

(b) The stated event is equivalent to 'at most 6 successes in 9 trials', that is, $X \leq 6$ where X denotes the number of successes in 9 trials. Using the binomial table with $n = 9, p = .7$ we find the required probability

$$P[X \leq 6] = .537$$

5.73 Mean $= np, \quad sd = \sqrt{npq}$

(a) Mean $= 18 \times .5 = 9, \quad sd = \sqrt{18 \times .5 \times .5} = 2.121$

(b) Mean $= 25 \times .2 = 5, \quad sd = \sqrt{25 \times .2 \times .8} = 2.0$

(c) Mean $= 25 \times .8 = 20, sd = \sqrt{25 \times .2 \times .8} = 2.0.$

5.75 Denote X = no. college seniors in support of increased funding, in a random sample of 20 seniors. Then X has the binomial distribution with $n = 20, p = .2$. We find $P[X \leq 3] = .411$.

5.77 Identify S : survival beyond 5 years

$$X = \text{no. patients surviving beyond 5 years, in a random sample of 19 patients.}$$

Then, X has the binomial distribution with $n = 19, p = .8$

(a) $P[X = 14] = .327 - .163 = .164$

(b) $P[X = 19 - 6] = P[X = 13] = .163 - .068 = .095$

(c) $P[9 \leq X \leq 13] = P[X \leq 13] - P[X \leq 8]$

$$= .163 - .000 = .163$$

5.79 $n = 545, \ p = .0785, \ q = .9215$

mean $= np = 42.7825$

$sd = \sqrt{npq} = 6.2789.$

5.81 (a) X has the binomial distribution with $n = 40$ and

$$p = P[\text{ Allergy present}] = .16 + .09 = .25$$

Therefore, $E(X) = 40 \times .25 = 10$

$$sd(X) = \sqrt{40 \times .25 \times .75} = 2.739$$

(b) Y has the binomial distribution with $n = 40$ and

$$p = P[\text{ Allergy present—Male}] = \frac{.16}{.16 + .36} = \frac{16}{52}$$

Therefore, $E(Y) = 40 \times \frac{16}{52} = 12.308$

$$sd(Y) = \sqrt{40 \times \frac{16}{52} \times \frac{36}{52}} = 2.919$$

(c) Z has the binomial distribution with $n = 40$ and

$$p = P[\text{ Allergy absent—Female}] = \frac{.39}{.09 + .39} = \frac{39}{48}$$

Therefore, $E(Z) = 40 \times \frac{39}{48} = 3.25$

$$sd(Z) = \sqrt{40 \times \frac{39}{48} \times \frac{9}{48}} = 2.469$$

5.83 Minitab binomial $n = 12$, $p = .67$

    (a) $P[X \leq 8] = 0.5973$

        $P[X = 8] = 0.2384$

    (b) $n = 35$, $p = .43$, $P[10 \leq X \leq 15] = 0.5372$

5.85 (a) From the example, the center line is at $p_0 = .4$, the lower control limit is .07 and the upper limit is .73.

    (b) The corresponding proportions .6, .5, .75, .55, and .8 are graphed below

    (c) The proportions .75 and .8, for the third and fifth weeks, are out of control.

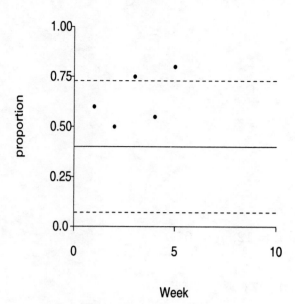

5.87 (a) From Exercise 5.86, the lower limit is .165 and the upper limit is .835. Since,

$$\frac{x}{20} \leq .165 \quad \text{only if} \quad x \leq 20 \times .165 = 3.3$$

and

$$\frac{x}{20} \geq .835 \quad \text{only if} \quad x \geq 20 \times .835 = 16.7$$

so the unusual values are 0,1, 2, 3 and 17, 18, 19, 20.

(b) When $n = 20, \quad p = .5,$

$$P[X \leq 3 \ \text{or} \ X \geq 17] = P[X \leq 3] + (1 - P[X \leq 16])$$

$$= .001 + (1 - .999) = .002$$

5.89 (a) Possible values of X are $-3, -1, 1, 3$.

(b) $X = -3: \quad TTT$

$\quad X = -1: \quad HTT, THT, TTH$

$\quad X = \quad 1: \quad HHT, HTH, THH$

$\quad X = \quad 3: \quad HHH$

*5.91 (a) Considering the selection of the box and the possible number of defectives in the sample, the elementary outcomes are

$$\begin{array}{ll} (\text{Box 1,} \quad 0D), & (\text{Box 2,} \quad 0D) \\ (\text{Box 1,} \quad 1D), & (\text{Box 2,} \quad 1D) \\ (\text{Box 1,} \quad 2D), & (\text{Box 2,} \quad 2D) \end{array}$$

Using conditional probability and the multiplication law, we calculate

$$P[\text{Box 1}, 0D] = P(\text{Box 1})P(0D|\text{Box 1})$$

$$= \frac{1}{2} \times \frac{\binom{15}{2}}{\binom{20}{2}} \qquad \text{because Box 1 contains 20 articles of which 15 are good and 5 are defective.}$$

$$= \frac{1}{2} \times \frac{105}{190} = .2763$$

In the same manner, we obtain

$$P[\text{Box 1, } 1D] = \frac{1}{2} \times \frac{\binom{5}{1}\binom{15}{1}}{\binom{20}{2}} = \frac{1}{2} \times \frac{75}{190} = .1974$$

$$P[\text{Box 1, } 2D] = \frac{1}{2} \times \frac{\binom{5}{2}}{\binom{20}{2}} = \frac{1}{2} \times \frac{10}{190} = .0263$$

$$P[\text{Box 2, } 0D] = \frac{1}{2} \times \frac{\binom{24}{2}}{\binom{30}{2}} = \frac{1}{2} \times \frac{276}{435} = .3172$$

$$P[\text{Box 2, } 1D] = \frac{1}{2} \times \frac{\binom{6}{1}\binom{24}{1}}{\binom{30}{2}} = \frac{1}{2} \times \frac{144}{435} = .1655$$

$$P[\text{Box 2}, 2D] = \tfrac{1}{2} \times \frac{\binom{6}{2}}{\binom{30}{2}} = \tfrac{1}{2} \times \tfrac{15}{435} = .0172$$

(b) The random variable Y, the number of defectives in the sample, can have the values 0, 1, 2, and the probabilities are readily obtained by adding the probabilities of the relevant elementary outcomes. For instance,

$$P[Y = 0] \; = P[\text{Box } 1, 0D] + P[\text{Box } 2, 0D]$$
$$= .2763 + .3172 = .5935$$

The probability distribution of Y is

y	$f(y)$
0	.5935
1	.3629
2	.0435
Total	.9999 (rounding error)

*5.93 (a) We denote the results of the successive matches by a string of letters A and B indicating the winner of each individual set. For instance, $AABA$ stands for the outcome that A wins the first two sets, B wins the third and A wins the fourth. In this case, the game stops at the fourth set so $X = 4$. Listed below are the possible values of X and the corresponding elementary outcomes:

$X = 3$: AAA,
$\quad\quad\quad$ BBB
$X = 4$: $AABA, ABAA, BAAA$,
$\quad\quad\quad$ $BBAB, BABB, ABBB$
$X = 5$: $AABBA, ABABA, ABBAA, BAABA, BABAA, BBAAA$,
$\quad\quad\quad$ $BBAAB, BABAB, BAABB, ABBAB, ABABB, AABBB$

(Note: In each case, the outcomes in the first line correspond to A being the winner and those in the second line correspond to B being the winner. Once the first line is completed, the second line can be readily obtained by interchanging the letters A and B).

(b) To calculate the probabilities we use the facts that $P(A) = .4, P(B) = .6$, and the results of the different sets are independent.

$P[X = 3] = (.4)^3 + (.6)^3 = .280$

$P[X = 4] = 3(.4)^3(.6) + 3(.6)^3(.4) = .3744$

$P[X = 5] = 6(.4)^3(.6)^2 + 6(.6)^3(.4)^2 = .3456$

The probability distribution of X is

x	$f(x)$
3	.2800
4	.3744
5	.3456
Total	1.0000

5.95

$$\begin{array}{|ll|} \hline 4 & \text{defective} \\ 8 & \text{good} \\ \hline 12 & \end{array} \rightarrow 3 \text{ sampled}$$

The possible values of X are 0, 1, 2, 3. The number of possible samples of 3 bulbs from 12 is

$$\left(\begin{array}{c} 12 \\ 3 \end{array} \right) = \frac{12 \times 11 \times 10}{3 \times 2 \times 1} = 220$$

These are all equally likely.

The probabilities of the various values of X are:

$$P[X = 0] = \frac{\binom{8}{3}}{\binom{12}{3}} = \frac{56}{220}$$

$$P[X = 1] = \frac{\binom{4}{1} \times \binom{8}{2}}{\binom{12}{3}} = \frac{4 \times 28}{220} = \frac{112}{220}$$

$$P[X = 2] = \frac{\binom{4}{2} \times \binom{8}{1}}{\binom{12}{3}} = \frac{6 \times 8}{220} = \frac{48}{220}$$

$$P[X = 3] = \frac{\binom{4}{3}}{\binom{12}{3}} = \frac{4}{220}$$

For instance, $X = 1$ corresponds to the occurrence of 1 defective and 2 good bulbs in the sample. The number of possible samples that have 1 defective and 2 good bulbs is $\binom{4}{1} \times \binom{8}{2}$.

x	$f(x)$
0	.255
1	.509
2	.218
3	.018
Total	1.000

5.97 (a) & (b)We calculate $\mu = \Sigma x_i f(x_i)$ and calculate σ^2 by the alternative formula

$$\sigma^2 = \Sigma x_i^2 f(x_i) - \mu^2.$$

x	$f(x)$	$xf(x)$	$x^2 f(x)$
2	.1	.2	.4
3	.3	.9	2.7
4	.3	1.2	4.8
5	.2	1.0	5.0
6	.1	.6	3.6
		3.9	16.5

$E(X) = \mu = 3.9$

$\sigma^2 = 16.5 - (3.9)^2 = 1.29,$ and $sd(X) = \sigma = \sqrt{1.29} = 1.14$

(c) The probability histogram of X with $\mu = E(X)$ located is given in Figure 5.8.

5.99 (a) The probability that the student will get either of the two winning tickets is $\frac{2}{1000} = .002$.

(b) Consider X = dollar amount of the student's winning. The random variable X can have the values 0 or 200 with probabilities .998 and .002, respectively.

x	$f(x)$	$xf(x)$
0	.998	0
200	.002	.4
		$.4 = E(X)$

Considering now the purchase price of $1, the student's expected gain =
$.40 − $1 = −$.60, that is, expected loss = $.60.

5.101 (a) $P[X < 3] = f(0) + f(1) + f(2) = .05 + .1 + .15 = .30.$

   (b)

x	$f(x)$	$xf(x)$	$x^2f(x)$
0	.05	0	0
1	.10	.10	.10
2	.15	.30	.60
3	.35	1.05	3.15
4	.20	.80	3.20
5	.15	.75	3.75
Total		$3.00 = \mu$	10.80

$E(X) = 3.00$

$\sigma^2 = \Sigma x^2 f(x) - \mu^2 = 10.80 - (3)^2 = 1.80$

$\sigma = \sqrt{1.80} = 1.342$

5.103 (a) Let A, B, C denote the correct names. Listed below are the elementary outcomes, that is, the possible assignments of names, and the corresponding values of $X =$ no. matches.

	Correct names			Value of
	A	B	C	X
	A	B	C	3
	A	C	B	1
Possible	B	A	C	1
assignments	B	C	A	0
	C	A	B	0
	C	B	A	1

(b) & (c) All 6 elementary outcomes are equally likely. Therefore,

$$P[X = 0] = \tfrac{2}{6}, \qquad P[X = 1] = \tfrac{3}{6}, \qquad P[X = 3] = \tfrac{1}{6}.$$

The probability distribution is presented below along with the calculation of its expectation.

x	$f(x)$	$xf(x)$
0	2/6	0
1	3/6	3/6
3	1/6	3/6
Total		$1 = E(X)$

5.105   (a) & (b) The possible values of X are:

$X = -15$ if he loses all three times

$X = 5 - 5 - 5 = -5$ if he loses twice and wins once

$X = 5 + 5 - 5 = 5$ if he loses once and wins twice

$X = 5 + 5 + 5 = 15$ if he wins all three times.

Using the symbol W for win and L for loss at each play we list the elementary outcomes. In assigning the probabilities, note that at each play $P(W) = \frac{18}{38}$ and $P(L) = \frac{20}{38}$ because 18 out of 38 slots are red, and 20 are not red. Also the outcomes at different plays are independent.

$X = -15$	$X = -5$	$X = 5$	$X = 15$
LLL	WLL	WWL	WWW
	LWL	WLW	
	LLW	LWW	

$P(LLL) = \left(\frac{20}{38} \right)^3 = .1458$ so $P[X = -15] = .1458$

$P(WLL) = \frac{18}{38} \times \frac{20}{38} \times \frac{20}{38}$, and the same result holds for LWL and LLW.

Therefore, $P[X = -5] = 3 \times \frac{18}{38} \times \left(\frac{20}{38} \right)^2 = .3936$

Similarly, $P[X = 5] = 3 \times \left(\frac{18}{38} \right)^2 \times \frac{20}{38} = .3543$

$P[X = 15] = \left(\frac{18}{38} \right)^3 = .1063$

Probability distribution of X and calculation of expectation

x	$f(x)$	$xf(x)$
-15	.1458	-2.187
-5	.3936	-1.968
5	.3543	1.7715
15	.1063	1.5945
Total	1.0000	$-.789$

$E(X) = -\$.79$

(c) No. At any play, betting on red or black are probabilistically equivalent because in either case $P(W) = \frac{18}{38}$. Also, different plays are independent.

5.107 In solving Exercise 5.11, we already obtained the probability distribution of $(X_1 + X_2)$, the total number of dots resulting from two tosses of a fair die.

Here we are concerned with the new random variable $\overline{X} = \dfrac{X_1 + X_2}{2}$. The distinct values of $X_1 + X_2$ lead to the distinct values of $\overline{X}$:

$$
\begin{array}{lccccccc}
X_1 + X_2 \text{ values}: & 2 & 3 & 4 & 5 & \ldots & 11 & 12 \\
\overline{X} \text{ values}: & 1 & 1.5 & 2 & 2.5 & \ldots & 5.5 & 6
\end{array}
$$

The probability distribution of $\bar{X}$ is

$\overline{x}$	$f(\overline{x})$
1	1/36
1.5	2/36
2	3/36
2.5	4/36
3	5/36
3.5	6/36
4	5/36
4.5	4/36
5	3/36
5.5	2/36
6	1/36
Total	1

5.109 (a)

x	$f(x)$	$F(x)$
1	.07	.07
2	.12	.19
3	.25	.44
4	.28	.72
5	.18	.90
6	.10	1.00

5.111

x	$f(x)$	$xf(x)$	$s^2f(x)$
2	2/15	4/15	8/15
3	4/15	12/15	36/15
4	6/15	24/15	96/15
5	3/15	15/15	75/15
total	1	55/15	215/15

$$\mu = \frac{55}{15} = 3.667$$

$$\sigma^2 = \frac{215}{15} - \left(\frac{55}{15}\right)^2 = .889$$

$$\sigma = \sqrt{.899} = .943$$

5.113 (a) Although there are two possible outcomes for each trial, the model of Bernoulli trials is not appropriate. The independence assumption is questionable.

(b) Bernoulli model is plausible.

(c) Not Bernoulli trials because time has more than two possible outcomes.

(d) There are two possible outcomes but the condition of independence does not hold. Clear or cloudy condition often lasts over several days.

(e) Bernoulli model is plausible.

5.115 Denoting male by M and female by F, we calculate

$$P(FFM) = .5 \times .5 \times .5 = .125.$$

5.117 Let $S =$ hitting a red light. We have $P(S) = .7$ and $P(F) = .3$.

    (a) $P(SS) = .7 \times .7 = .49$

    (b) $P(SSS) = (.7)^3 = .343$

    (c) $P(SSF) + P(SFS) + P(FSS) = 3 \times (.7)^2 \times .3 = .441$.

5.119 $P(FF) = q^2 = \frac{4}{49}$ so $q = \sqrt{\frac{4}{49}} = \frac{2}{7}$ and $p = 1 - q = \frac{5}{7}$

    $P(SSF) = p^2 q = \left(\frac{5}{7} \right)^2 \frac{2}{7} = \frac{50}{343} = .146$.

5.121 (a) X has the binomial distribution with $n = 6$ and $p = .4$.

(b) Using the binomial table, we find

$$P[X \leq 3] = .821, P[X = 0] = .047.$$

$$E(X) = np = 6 \times .4 = 2.4 \quad \text{persons}$$

5.123   (a) $P[SSSFFFF] = p^3 q^4 = .008$

$P[\text{3 successes in 7 trials}] = \left(\begin{array}{c} 7 \\ 3 \end{array} \right) p^3 q^4 = 35 p^3 q^4 = 35 \times .008 = .28$

(b) $P[\text{2 successes in 7 trials}] = \left(\begin{array}{c} 7 \\ 2 \end{array} \right) p^2 q^5 = 21 p^2 q^5 = .0336$

$P[FFFFFSS] = q^5 p^2 = \dfrac{.0336}{21} = .0016.$

5.125 Employing the binomial model with $n = 20$ and $p = .7$, we find $P[X \leq 10] = .048$. The probability of the observed result 10, or a more extreme result, is so small that we would doubt the claim that $p = .7$. Not as many students as claimed support the paper's view.

5.127 We use the binomial table for $n = 14, p = .4$.

(a) $P[3 \leq X \leq 9] = P[X \leq 9] - P[X \leq 2] = .982 - .040 = .942$

(b) $P[3 < X \leq 9] = P[X \leq 9] - P[X \leq 3] = .982 - .124 = .858$

(c) $P[3 < X < 9] = P[4 \leq X \leq 8]$

$$= P[X \leq 8] - P[X \leq 3] = .942 - .124 = .818$$

(d) $E(X) = 14 \times .4 = 5.6$

(e) $sd(X) = \sqrt{14 \times .4 \times .6} = 1.833$.

5.129 $P[\text{1 success in } n \text{ trials}] = \begin{pmatrix} n \\ 1 \end{pmatrix} pq^{n-1} = npq^{n-1}$

$P[\text{0 success in } n \text{ trials}] = q^n$

One success is more probable than 0 success if $npq^{n-1} > q^n$ or

$$n > \frac{q^n}{pq^{n-1}} = \frac{q}{p} = \frac{.85}{.15} = 5.67$$

The smallest n (integer) that satisfies $n > 5.67$ is 6.

5.131 (a) $n = 12$

p	.1	.2	.3	.4	.5
$P[X \le 3]$	.974	.795	.493	.225	.073

(b) $n = 18$

p	.1	.2	.3	.4	.5
$P[X \le 3]$	.902	.501	.165	.033	.004

5.133 Let $X =$No. of persons (out of 20) who feel the system is adequate. Since the city population is large, the binomial distribution is appropriate for X. We have $n = 20$ and $p = .3$ so

$$P[X \geq 10] = 1 - P[X \leq 9] = 1 - .952 = .048$$

$$P[X = 10] = P[X \leq 10] - P[X \leq 9] = .983 - .952 = .031.$$

5.135 $E(X) = np = 16 \times .2 = 3.2$

$$sd(X) = \sqrt{npq} = \sqrt{16 \times .2 \times .8} = 1.6.$$

*5.137 (a) $P[X = 0] = f(0) = e^{-3}\dfrac{(3)^0}{0!} = e^{-3} = .05$

       (b) $P[X = 1] = f(1) = e^{-3}\dfrac{(3)^1}{1!} = e^{-3} \times 3 = .15.$

Chapter 6

THE NORMAL DISTRIBUTION

6.1 (a) The function is non-negative and the area of the rectangle is $.5 \times 2 = 1$. It is a probability density function.

   (b) Since $f(x)$ takes negative values over the interval from 1 to 2, it is not a probability density function.

   (c) The function is non-negative and the area of the triangle is $\frac{1}{2} \times base \times height = \frac{1}{2} \times 2 \times 1 = 1$. It is a probability density function.

   (d) The function is non-negative but the area of the rectangle is $1 \times 2 = 2$ so it is not a probability density function.

6.3 From the picture of $f(x)$ we see that the interval 1.5 to 2 has a larger area under $f(x)$ than the interval 0 to .5. Consequently, $P[1.5 < X < 2]$ is the larger of the two probabilities.

225

6.5 For an arbitrary point x in the interval 0 to 2, we find that

$$P[0 < X < x] = \text{Area of the triangle over the base} \quad (0, x)$$

$$= \frac{1}{2} \cdot x \cdot \frac{x}{2} = \frac{x^2}{4}$$

The median is the value of x for which the cumulative probability $x^2/4 = .5$ so $x = \sqrt{2} = 1.414$. Similarly, $x = \sqrt{1} = 1$ is Q_1 and $x = \sqrt{3} = 1.73$ is Q_3.

6.7 (a) The median is the time such that there is an equal probability of being earlier or of being later. Consequently, the median is later than 1:20 p.m.

(b) No. The mean could be larger or smaller than the median depending on the distribution. The mean is often larger than the median when the distribution has a long tail to the right.

6.9 The standardized variable is

$$Z = \frac{X - \mu}{\sigma}$$

   (a) $Z = (X - 12)/4$

   (b) $Z = (X - 21)/9$

   (c) $Z = (X - 121)/\sqrt{25} = (X - 121)/5$

6.11 (a) $Z = (X - 7)/2$

   (b) $Z = (X - 250)/8$

   (c) $Z = (X - 777)/9$

6.13 We use Appendix Table 3 which gives the area under the standard normal curve to the left of a $z-$ value.

(a) $P[Z < .83] = .7967$

(b) $P[Z < 1.03] = .8485$

(c) $P[Z < -1.03] = .1515$

(d) $P[Z < -1.35] = .0885$

6.15 The area to the right of a $z-$ value $= 1 -$ Area to the left of the $z-$ value.

   (a) $P[Z > .83] = 1 - P[Z < .83] = 1 - .7967 = .2033$

   (b) $1 - P[Z < 2.83] = 1 - .9977 = .0023$

   (c) $1 - P[Z < -1.23] = 1 - .1093 = .8907$

   (d) $z = 1.635$ is the mid-point between $z = 1.63$ and $z = 1.64$. From the
        normal table, we find

        $$
        \begin{array}{ll}
        P[Z < 1.64] & = .9495 \\
        P[Z < 1.63] & = .9484 \\
        \hline
        \text{difference} & = .0011
        \end{array}
        $$

The mid-point between .9495 and .9484 is

$$.9484 + \frac{1}{2}(.0011) = .9490$$

so $P[Z \leq 1.635] = .9490$ and $P[Z > 1.635] = 1 - .9490 = .0510$.

6.17   (a) $P[-.44 < Z < .44] = P[Z < .44] - P[Z < -.44]$
        $= .6700 - .3300 = .3400.$

(b) $P[-1.33 < Z < 1.33] = P[Z < 1.33] - P[Z < -1.33]$

$= .9082 - .0918 = .8164.$

(c) $P[.40 < Z < 2.03] = P[Z < 2.03] - P[Z < .40]$

$= .9788 - .6554 = .3234.$

(d) We find

$$
\begin{array}{ll}
P[Z < 1.41] & = .9207 \\
\underline{P[Z < 1.40]} & \underline{= .9192} \\
\text{difference} & = .0015
\end{array}
$$

Therefore,

$$P[Z < 1.405] = .9192 + \frac{1}{2}(.0015) = .9200$$

Also,

$$
\begin{array}{ll}
P[Z < 2.31] & = .9896 \\
\underline{P[Z < 2.30]} & \underline{= .9893} \\
\text{difference} & = .0003
\end{array}
$$

so

$$P[Z < 2.306] = .9893 + .6 \times (.0003) = .9895$$

Finally,

$$P[1.405 < Z < 2.306] = .9895 - .9200 = .0695$$

6.19 (a) We are to find the z-value for which the area to the left is .20 . From the

normal table , we find

$$P[Z < -.84] \quad = .2005$$
$$P[Z < -.85] \quad = .1977$$
$$\overline{\text{difference} \quad = .0028}$$

Since $.2005 - .20 = .0005$, the required z−value is

$$-.84 - (.01) \times \frac{.0005}{.0028} = -.842$$

(b) We are to find the z−value for which the area to the left is $1 - .125 = .875$

so $z = 1.15$.

(c) By the symmetry of the normal curve, we have

$$2P[Z < -z] = 1 - P[-z < Z < z]$$

$$= 1 - .668 = .332$$

or $P[Z < -z] = .166$ so $z = .97$.

(d) We are to find the z−value for which

$$.888 = P[z < Z < 2.0] = P[Z < 2.0] - P[Z < z] = .9772 - P[Z < z]$$

or $P[Z < z] = .0892$. We find

$$P[Z < -1.34] \quad = .0901$$
$$P[Z < -1.35] \quad = .0885$$
$$\overline{\text{difference} \quad = .0016}$$

the required z−value is

$$-1.34 - (.01) \times \frac{.0009}{.0016} = -1.34 - .006 = -1.346$$

6.21 (a) $P[Z < -.93] = .1762$ so the $z-$value is $-.93$.

(b) We are to find the $z-$value for which the area to the left is $1 - .10 = .90$. From the normal table, we find

$$P[Z < 1.29] = .9015$$
$$\underline{P[Z < 1.28] = .8997}$$
$$\text{difference} = .0018$$

since $.8997$ is nearly $.90$ we could take $z = 1.28$. More accurately, the required $z-$value is

$$1.28 + (.01) \times \frac{.0003}{.0018} = 1.28 + .0017 = 1.2817 \quad \text{or} \quad 1.282(\text{ rounded})$$

(c) Since $2P[Z < -z] = 1 - .954 = .046$, we require that $P[Z < -z] = \frac{1}{2}(.046) = .023$. From the normal table,

$$P[Z < -1.99] = .0233$$
$$\underline{P[Z < -2.00] = .0228}$$
$$\text{difference} = .0005$$

so

$$-1.99 - (.01) \times \frac{.0003}{.0005} = -1.996$$

is the required $z-$ value.

(d) We require that

$$P[Z < z] = .50 + P[Z < -.6] = .50 + .2743 = .7743$$

Scanning the normal table, we find

$$
\begin{array}{ll}
P[Z < .76] & = .7764 \\
P[Z < .75] & = .7734 \\
\hline
\text{difference} & = .0030
\end{array}
$$

The probability .7743 is about a third of way so we could use .753. More accurately , the $z-$ value is

$$.75 + (.01) \times \frac{.0009}{.0030} = .753$$

6.23   (a) $P[Z < .33] = .6293$

(b) We find $P[Z < -.44] = .3300$ Therefore, the 33-th percentile is $-.44$.

(c) $P[Z < .70] = .7580$

(d) We find

$$
\begin{array}{ll}
P[Z < .53] & = .7019 \\
P[Z < .52] & = .6985 \\
\hline
\text{difference} & = .0034
\end{array}
$$

Therefore, the 70-th percentile is

$$.52 + (.01) \times \frac{.0015}{.0034} = .524$$

6.25 The standardized variable is $Z = \dfrac{X - 60}{4}$.

(a) For $x = 55$, we have $z = (55 - 60/4) = -1.25$ so

$$P[X < 55] = P[Z < -1.25] = .1056$$

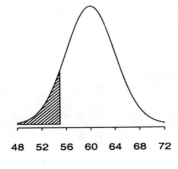

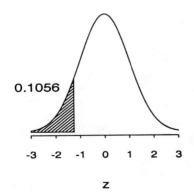

(b)

$$P[X \leq 67] = P[Z \leq 1.75] = .9599$$

(c)

$$P[X > 66] = P[Z > 1.50] = 1 - P[Z \leq 1.50] = 1 - .9332 = .0668$$

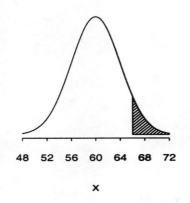 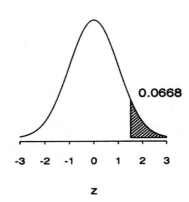

(d)

$$P[X > 51] = P[Z > -2.25] = 1 - .0122 = .9878$$

(e)

$$P[53 \leq X \leq 69] = P[\frac{53 - 60}{4} \leq Z \leq \frac{69 - 60}{4}] = P[-1.75 \leq Z \leq 2.25]$$

$$= P[Z < 2.25] - P[Z < -1.75] = .9878 - .0401 = .9477$$

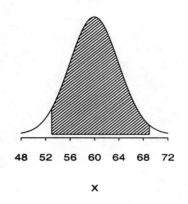

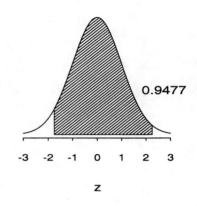

(f)

$$P[61 < X < 64] = P[.25 < Z < 1.00] = .8413 - .5987 = .2426$$

6.27 The standardized variable is $Z = \frac{X - 120}{5}$.

(a) From the normal table we find $P[Z < 1.96] = .975$

Therefore, $\frac{b - 120}{5} = 1.96$ so $b = 120 + 5(1.96) = 129.8$

(b) Here $P[X < b] = 1 - .025 = .975$

As calculated in part (a), we have $b = 129.8$

(c) Since $P[Z < -.51] = .305$, we must have

$$\frac{b - 120}{5} = -.51 \text{ or } b = 120 + 5(-.51) = 117.45.$$

6.29 Let X denote the score of a randomly selected student. We have $\mu = 500$ and $\sigma = 100$ so the standardized variable is $Z = \frac{X - 500}{100}$.

(a) $P[X > 650] = P[Z > \frac{650 - 500}{100}] = P[Z > 1.5]$

$$= 1 - .9332 = .0668$$

(b) $P[X < 250] = P[Z < -2.5] = .0062$

(c) $x = 325$ gives $z = \frac{325 - 500}{100} = -1.75$

$x = 675$ gives $z = \frac{675 - 500}{100} = 1.75$

$P[325 < X < 675] = P[-1.75 < Z < 1.75]$

$$= .9599 - .0401 = .9198.$$

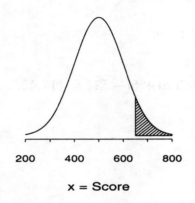

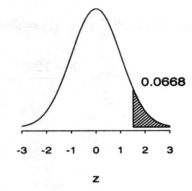

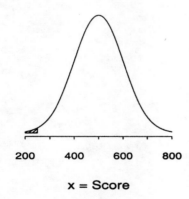

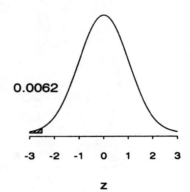

6.31 The z value for 32.5 inches is $(32.5 - 34.5)/1.3 = -1.538$ and that for 36.5 inches is $(36.5 - 34.5)/1.3 = 1.538$ so

$$P[32.5 < X < 36.5] = P[-1.538 < Z < 1.538] = .9378 - .0623 = .8755$$

6.33 The z value for 4 ounces is $(4 - 5)/1.2 = -.833$ so

$$P[X < 4] = P[Z < -.833] = .2024$$

6.35 From the standard normal table, we require that the central probability

$P[|Z| \leq z] = .80 = 1 - 2P[Z \leq -z]$ or $P[Z \leq -z] = .10$. Interpolating in the table, we find that $P[Z \leq -1.282] = .100$ so 1.282 is the z value. Since

$$Z = \frac{X - 69.0}{2.8} \quad or \quad X = 2.8Z + 69.0$$

the shortest person is $2.8(-1.282) + 69 = 65.4$ inches and the tallest is $2.8\,(1.282) + 69.0 = 72.6$ inches.

6.37 The arrival time X is distributed as $N(17, 3)$.

(a) (i) $P[X > 22] = P[Z > \frac{22 - 17}{3}]$

$= P[Z > 1.67] = 1 - .9525 = .0475$

(ii) The z values corresponding to $x = 13$ and $x = 21$ are

$\frac{13 - 17}{3} = -1.33$ and $\frac{21 - 17}{3} = 1.33$, respectively.

$P[13 < X < 21] = P[-1.33 < Z < 1.33]$

$= .9082 - .0918 = .8164$

(iii) $\frac{15.5 - 17}{3} = -.5$, $\frac{18.5 - 17}{3} = .5$

$P[15.5 < X < 18.5] = P[-.5 < Z < .5]$

$= .6915 - .3085 = .3830$

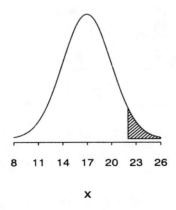
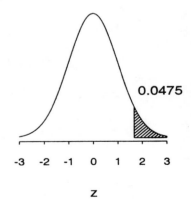

(b) The probability density curve of a normal distribution peaks at the mean. Therefore, the 1-minute interval that has the highest probability is one that is centered at $\mu = 17$ and has length 1, that is, the interval 16.5 to 17.5.

6.39   (a) We use the binomial table (Appendix Table 2) for $n = 25$ and $p = .6$.

(i) $P[X = 17] = P[X \leq 17] - P[X \leq 16] = .846 - .726 = .120$

(ii) $P[11 \leq X \leq 18] = P[X \leq 18] - P[X \leq 10]$

$$= .926 - .034 = .892$$

(iii) $P[11 < X < 18] = P[X \leq 17] - P[X \leq 11]$

$$= .846 - .078 = .768$$

(b) With $n = 25$ and $p = .6$, we calculate

$$np = 15 \text{ and } \sqrt{npq} = \sqrt{25 \times .6 \times .4} = 2.45.$$

We consider X to be normally distributed with mean $= 15$ and $sd = 2.45$ so

$$Z = \frac{X - 15}{2.45} \approx N(0, 1)$$

(i) Using the continuity correction, we calculate the normal probability assigned to the interval 16.5 to 17.5.

$$P[16.5 < X < 17.5] = P[\frac{16.5 - 15}{2.45} < Z < \frac{17.5 - 15}{2.45}]$$

$$= P[.612 < Z < 1.020]$$

$$= .8461 - .7298 = .1163.$$

(ii) With continuity correction, the relevant interval is 10.5 to 18.5

$$P[10.5 < X < 18.5] = P[\frac{10.5 - 15}{2.45} < Z < \frac{18.5 - 15}{2.45}]$$

$$= P[-1.837 < Z < 1.429]$$

$$= .9235 - .0331 = .8904$$

(iii) Here the end points 11 and 18 are not included. The continuity correction leads to the interval 11.5 to 17.5.

$$P[11.5 < X < 17.5] = P[-1.429 < Z < 1.020]$$

$$= .8461 - .0765 = .7696.$$

6.41 To use the normal approximation, we calculate

$$\text{mean} = np = 300 \times .25 = 75$$

$$sd = \sqrt{npq} = \sqrt{300 \times .25 \times .75} = 7.50$$

so $Z = \dfrac{X - 75}{7.50}$ is approximately $N(0,1)$.

(a) We calculate the normal probability of the interval 79.5 to 80.5.

$$P[X = 80] \approx P[\frac{79.5 - 75}{7.50} < Z < \frac{80.5 - 75}{7.50}]$$

$$= P[.60 < Z < .733] = .7683 - .7257 = .0426$$

(b) Using the continuity correction,

$$P[X \leq 65] \approx P[Z < \frac{65.5 - 75}{7.50}]$$

$$= P[Z < -1.267] = .1025$$

(c) $P[68 \leq X \leq 89] \approx P[\frac{67.5 - 75}{7.50} < Z < \frac{89.5 - 75}{7.50}]$

$$= P[-1.0 < Z < 1.933]$$

$$= .9734 - .1587 = .8147$$

6.43 (a) Normal approximation is appropriate because n is large and p not too close to 0 or 1.

(b) Not appropriate because p is too small, $np = 3$.

(c) Normal approximation is appropriate because n is large and p not too close to 0 or 1.

(d) Not appropriate because p is too close to 1, $n(1 - p) = 2.4$.

6.45 The standarized scale is $z = \dfrac{x - np}{\sqrt{np(1 - p)}}$

$\quad n = 5, \quad p = .4 : \quad z = (x - 2)/1.095$

$\quad n = 12, \quad p = .4 : \quad z = (x - 4.8)/1.697$

$\quad n = 25. \quad p = .4 : \quad z = (x - 10)/2.449.$

The probability histograms of binomial distribution for $p = .4$, $n = 5, 12$, and 25, and the corresponding z-scores are given in Figure 6.1.

6.47 Let X = no. of unemployed persons in a random sample of 300. Then distribution of X is binomial with $n = 300$ and $p = .079$, and a normal approximation is appropriate. Here

$$np = 300 \times .079 = 23.7$$

$$\sqrt{npq} = \sqrt{300 \times .079 \times .921} = 4.672$$

and $\qquad\qquad Z = \dfrac{X - 23.7}{4.672}$

(a) $P[X < 18] \approx P[Z < \dfrac{17.5 - 23.7}{4.672}] = P[Z < -1.327] = .0923$

(b) $P[X > 30] \approx P[Z > \dfrac{30.5 - 23.7}{4.672}] = P[Z > 1.455] = 1 - .9272 = .0728.$

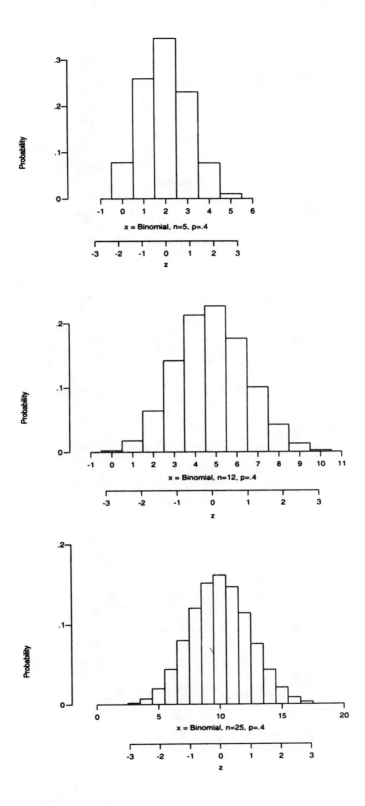

Figure 6.1. The binomial distribution for $p = .4$ and $n = 5, 12,$ and 25.

6.49 We assume the customers make purchases independently of one another. The
      binomial distribution applies. Since $n = 80$ and $p = .20$, the distribution of X
      = number who make a purchase, is approximately normal with

      mean $= 80 \times .20 = 16.0$ and $sd = \sqrt{80 \times .20 \times .80} = 3.578$

      Using the continuity correction, we approximate

$$P[X > 20] = P[Z > \frac{20.5 - 16.0}{3.578}] = 1 - P[Z < 1.258] = 1 - .8958 = .1042$$

6.51 Let Y be next weeks expenditure so

$$p = P[Y > 880] = P[Z > \frac{880 - 850}{40}] = P[Z > 0.75] = 1 - .7734 = .2266$$

Let X be the number of weeks, out of $n = 52$, where the expenses exceed 580 dollars. Then X has the binomial distribution with $n = 52$ and $p = .2266$. so to use the normal approximation, we calculate

$$\mu = np = 52 \times .2266 = 11.783 \quad \text{and} \quad \sigma = \sqrt{npq} = \sqrt{52 \times .2266 \times .7734} = 3.019$$

Using the continuity correction,

$$P[10 \leq X \leq 16] \approx P[\frac{9.5 - 11.783}{3.019} < Z \leq \frac{16.5 - 11.783}{3.019}]$$

$$= P[-.756 < Z < 1.562] = .9409 - .2248 = .7161.$$

6.53 We assume the residents make their decision to move independently of one another. The binomial distribution applies. Since $n = 100$ and $p = .32$, the distribution of $X =$ number who move , is approximately normal with

mean $= 100 \times .32 = 32 \quad$ and $\quad sd = \sqrt{100 \times .32 \times .68} = 4.665$

Using the continuity correction, we approximate

$$P[X \geq 39] \approx P[Z > \frac{38.5 - 32}{4.665}] = 1 - P[Z < 1.393] = 1 - .9182 = .0818$$

6.55 The probability density curve is a rectangle with height $= 1$.

(a) The point $x = .5$ divides the distribution into halves so median $= .5$.

(b) The points $x = .25, .5$ and $.75$ divide the distribution into quarters. Therefore, first quartile $= .25$, second quartile $= .5$, and third quartile $= .75$.

6.57 We could make a histogram using the first 100 heights and then another based on the first 500 heights. The histogram at the front of the chapter is based on 1,456 observations, a large number. From these original data of heights, one can construct a histogram with considerably smaller class intervals than given in the chapter. If this process is repeated with an even much larger sample, the jumps between consecutive rectangles will then dampen out, and the top of the histogram will approximate the shape of a smooth curve. This density curve would be non-negative and the area under the curve is one.

6.59 (a) Scanning the probabilities given in the normal table, we find

$P[Z < -1.38] = .0838$. Therefore $z = -1.38$.

(b) In the table we look for probabilities close to .047, and find

$$P[Z < -1.68] = .0465 \text{ and } P[Z < -1.67] = .0475$$

Interpolation gives $P[Z < -1.675] = .047$ so $z = -1.675$.

(c) Area to the left of z is $1 - .2611 = .7389$. From the table we get

$P[Z < .64] = .7389$ so $z = .64$.

(d) Area to the left so z is $1 - .12 = .88$. The table gives

$$P[Z < 1.17] = .8790 \quad \text{and} \quad P[Z < 1.18] = .8810$$

Since .88 is half way between these results, we interpolate $z = 1.175$.

6.61 (a) $P[Z > -.72] = 1 - P[Z < -.72] = 1 - .2358 = .7642$

(b) $P[-1.5 < Z < 1.5] = P[Z < 1.5] - P[Z < -1.5] = .9332 - .0668 = .8664$

(c) $P[|Z| > 2] = P[Z > 2] + P[Z < -2]$

$$= 2P[Z < -2] \quad \text{(by symmetry)}$$

$$= 2 \times .0228 = .0456$$

(d) $P[|Z| < 1] = P[-1 < Z < 1] = P[Z < 1] - P[Z < -1] = .8413 - .1587$

$\quad = .6826$

Alternatively, we can calculate

$P[|Z| > 1] = 2P[Z < -1] \qquad \text{(as in part (b))}$

$\qquad \qquad = 2 \times .1587 = .3174$

Hence $P[|Z| < 1] = 1 - .3174 = .6826$.

6.63 The standard normal variable is

$$Z = \frac{X - \mu}{\sigma} = \frac{X - 100}{8}$$

(a) $P[X < 107] = P[Z < \frac{107 - 100}{8}] = P[Z < .875] = .8092$

(b) $P[X < 97] \ = P[Z < -.375] = .3539$

(c) $P[X > 110] = P[Z > 1.25] = 1 - .8944 = .1056$

(d) $P[X > 90] \ = P[Z > -1.25] = 1 - .1056 = .8944$

(e) $P[95 < X < 106] \ = P[-.625 < Z < .75] = .7734 - .2660 = .5074$

(f) $P[103 < X < 114] = P[.375 < Z < 1.75] = .9599 - .6461 = .3138$

(g) $P[88 < X < 100]\ \ = P[-1.5 < Z < 0] = .5 - .0668 = .4332$

(h) $P[60 < X < 108]\ \ = P[-5 < Z < 1] = .8413 - 0 = .8413.$

6.65 The standardized variable is $Z = \dfrac{X - 497}{120}$.

(a) $P[X > 600] = P[Z > \dfrac{600 - 497}{120}] = P[Z > .858] = 1 - .8045 = .1955$

(b) We first find the 90th percentile of the standard normal distribution and then convert it to the x scale.

$$P[Z < 1.28] = .8997 \approx .90$$

The standardized score $z = 1.28$ corresponds to

$$x = 497 + 120(1.28) = 651$$

(c) $P[X < 400] = P[Z < -.808] = .2096.$

6.67 The strength X is distributed as $N(100, 8)$. The bonding will fail if $X < 98$. Its probability is

$$P[X < 98] = P[Z < \frac{98 - 100}{8}] = P[Z < -.25] = .4013.$$

6.69 $X =$ no. of days of trouble-free operation is normally distributed with $\mu = 530$ and $\sigma = 100$ so the standardized variable is $Z = \frac{X - 530}{100}$.

(a) Taking 365 days in a year, 2 years have 730 days.

$$P[X > 730] = P[Z > \frac{730 - 530}{100}] = P[Z > 2] = 1 - .9772 = .0228.$$

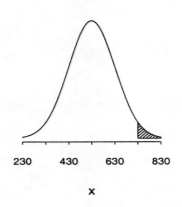 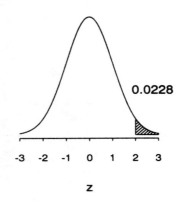

(b) Denoting d = no. of days of warranty, we require that $P[X < d] = .1$.

From the standard normal table, we find $P[Z < -1.28] = .10$ so

$$\frac{d - 530}{100} = -1.28 \text{ or } d = 530 + 100(-1.28) = 402$$

The warranty period can be set to 402 (or about 400) days.

6.71 (a) We have the relation $Y = a + bX$ with $a = -\frac{5}{9}(32)$ and $b = \frac{5}{9}$.

Since X has the mean $\mu = 102$ and sd $= \sigma = 4$, the mean and sd of Y are:

$$\text{mean} = a + b\mu = -\frac{5}{9}(32) + \frac{5}{9}(102) = 38.89$$

$$\text{sd} = |b|\sigma = \frac{5}{9}(4) = 2.222$$

Y has the normal distribution $N(38.89,\ 2.222)$.

(b) The standardized variable is $Z = \frac{Y - 38.89}{2.222}$.

$$P[35 \leq Y \leq 42] = P[\frac{35 - 38.89}{2.222} \leq Z \leq \frac{42 - 38.89}{2.222}]$$

$$= P[-1.75 < Z < 1.40] = .9192 - .0401 = .8791.$$

6.73 Let $X = $ no. of late payments out of 20,000 tax bills. It has a binomial distribution with $n = 20,000$ and $p = .07$. For a normal approximation, we calculate

$$np = 20,000 \times .07 = 1400 \ , \quad \sqrt{npq} = \sqrt{20,000 \times .07 \times .93} = 36.08.$$

The continuity correction is not important because the sd is fairly large.

(a) $P[X < 1350] \approx P[Z < \frac{1350 - 1400}{36.08}] = P[Z < -1.386] = .0829.$

(b) $P[X \geq 1480] \approx P[Z > \frac{1480 - 1400}{36.08}] = P[Z > 2.217] = 1 - .9867 = .0133$

6.75 (a) n is large and p not too close to 0 or 1 so the normal approximation is appropriate.

(b) n is not large and also p is very small so the normal approximation is not appropriate.

(c) Although n is large, p being too close to 1, the normal approximation is not appropriate.

6.77 The large values of volume are too large for the normal distribution to hold. They should be pulled down with a transformation.

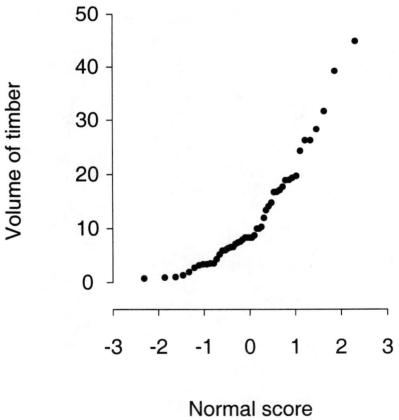

6.79 The normal-scores plot in Figure 6.3 is nearly straight.

6.81 Let X be the number of new words in the poem. We approximate the probability, that a new word will not be on the list, by $14,376/884,647 = .01625$. Then, we treat X as having a binomial distribution with $n = 429$ and $p = .01625$.

(a)

$$\text{expected number of new words} \quad = \quad \text{number of words} \times \text{probability}$$

$$= 429 \times .01625 = 6.97$$

(b) The standard deviation of X is $\sqrt{429 \times .01625 \times .98375} = 2.619$. Using the continuity correction, we approximate the binomial probability

$$P[X \geq 12] \approx P[Z \geq \frac{11.5 - 6.97}{2.619}] = P[Z \geq 1.73] = 1 - .9582 = .0418$$

(c) Using the continuity correction, we approximate the binomial probability

$$P[X \leq 2] \approx P[Z \leq \frac{2.5 - 6.97}{2.619}] = P[Z \leq -1.707] = .0439$$

(d)

$$P[2 < X < 12] \approx P[\frac{2.5 - 6.97}{2.619} \leq Z \leq \frac{11.5 - 6.97}{2.619}]$$

$$= P[Z \leq \frac{11.5 - 6.97}{2.619}] - P[Z \leq \frac{2.5 - 6.97}{2.619}]$$

$$= P[Z \leq 1.73] - P[Z \leq -1.707] =$$

$$.9582 - .0439 = .9143$$

The observed number 9 is very to the expected number 6.97. The difference $9 - 6.97 = 2.03$ is not large according to our probability calculation with shows that ± 4 words from 6.97 is not unusual.

Chapter 7

VARIATION IN REPEATED SAMPLES - SAMPLING DISTRIBUTIONS

7.1 (a) Statistic   (b) Statistic   (c) Parameter

   (d) Statistic   (e) Parameter

7.3 (a) Student's working full time would be more likely to take an evening course than students who do not work and so they would be over represented in the sample.

(b) A large majority of persons, at all income levels, spend more during the holiday season. The total amount spent and the types of purchases are often atypical. The survey would likely be misleading.

7.5 (a) All possible samples (x_1, x_2) and the corresponding $\bar{x}$ values are:

(x_1, x_2)	(3,3)	(3,5)	(3,7)	(5,3)	(5,5)	(5,7)	(7,3)	(7,5)	(7,7)
$\bar{x} = \dfrac{x_1 + x_2}{2}$	3	4	5	4	5	6	5	6	7

(b) The 9 possible samples are equally likely so each has the probability 1/9. The sampling distribution of $\bar{X}$ is obtained by listing the distinct values of $\bar{X}$ and the corresponding probabilities.

$\bar{x}$	Probability $f(\bar{x})$
3	1/9
4	2/9
5	3/9
6	2/9
7	1/9
Total	1

7.7 Not a random sample. The photographer would show his better pictures when trying to get a contract for wedding pictures.

7.9 We have $\mu = 79$ and $\sigma = 9$.

(a) For $n = 4$, $E(\bar{X}) = \mu = 79$ and $\mathrm{sd}(\bar{X}) = \dfrac{\sigma}{\sqrt{n}} = \dfrac{9}{\sqrt{4}} = 4.5$

(b) For $n = 25$, $E(\bar{X}) = \mu = 79$ and $\mathrm{sd}(\bar{X}) = \dfrac{9}{\sqrt{25}} = 1.8$

7.11 The population sd is $\sigma = 20$ so $\text{sd}(\bar{X}) = \frac{\sigma}{\sqrt{n}} = \frac{20}{\sqrt{n}}$

(a) $\text{sd}(\bar{X}) = \frac{20}{\sqrt{25}} = 4$

(b) $\text{sd}(\bar{X}) = \frac{20}{\sqrt{100}} = 2$

(c) $\text{sd}(\bar{X}) = \frac{20}{\sqrt{400}} = 1$

7.13 We first calculate the mean μ and sd σ of the population that corresponds to

X taking the values 3, 5, 7 with equal probabilities

Calculation of μ and σ

x	$f(x)$	$xf(x)$	$x^2 f(x)$
3	1/3	3/3	9/3
5	1/3	5/3	25/3
7	1/3	7/3	49/3
Total	1	15/3	83/3
		$= E(X)$	$= E(X^2)$

$\mu = \frac{15}{3} = 5$

$\sigma^2 = E(X^2) - \mu^2$

$= \frac{83}{3} - 25 = \frac{8}{3}$

$\sigma = \sqrt{8/3}$

For $n = 2$, we must have

$$E(\bar{X}) = \mu = 5 \text{ and } \text{sd}(\bar{X}) = \frac{\sigma}{\sqrt{2}} = \sqrt{\frac{8}{6}} = \sqrt{\frac{4}{3}}$$

We verify these by calculation from the distribution of $\bar{X}$:

Calculation of $E(\bar{X})$ and sd$(\bar{X})$

$\bar{x}$	$f(\bar{x})$	$\bar{x}f(\bar{x})$	$\bar{x}^2 f(\bar{x})$
3	1/9	3/9	9/9
4	2/9	8/9	32/9
5	3/9	15/9	75/9
6	2/9	12/9	72/9
7	1/9	7/9	49/9
Total	1	45/9	237/9

$$E(\bar{X}) = \frac{45}{9} = 5 = \mu$$

$$\text{Var}(\bar{X}) = \frac{237}{9} - 5^2 = \frac{12}{9} = \frac{4}{3}$$

$$\text{sd}(\bar{X}) = \sqrt{4/3}$$

7.15   (a) $E(\bar{X}) = \mu = 24$

(b) $\text{sd}(\bar{X}) = \frac{\sigma}{\sqrt{n}} = \frac{5}{\sqrt{4}} = 2.5$

(c) Since the population distribution is normal, the sample mean $\bar{X}$ has the normal distribution with mean $= 24$ and sd $= 2.5$.

7.17   The population mean and sd are $\mu = 94$ and $\sigma = 22$. Therefore

$$E(\bar{X}) = 94 \text{ and } \text{sd}(\bar{X}) = \frac{\sigma}{\sqrt{n}} = \frac{22}{\sqrt{5}} = 9.839$$

Since the population distribution is normal, $\bar{X}$ has the normal distribution with mean $= 94$ and sd $= 9.839$. The standardized variable is $Z = \frac{\bar{X} - 94}{9.839}$ and we need to obtain

$$P[\bar{X} > 100] = P[Z > \frac{100 - 94}{9.839}] = P[Z > .61] = .2709$$

7.19 (a) From the normal table, $P[Z < -2.054] = .020$. If μ_F is the mean fill rate, $Z = \frac{\bar{X} - \mu_F}{.0407}$ is the standardized variable and

$$P[\bar{X} < 16.0] = P[Z < \frac{16.0 - \mu_F}{.0407}] = .020$$

if

$$\frac{16.0 - \mu_F}{.0407} = -2.054 \quad \text{or} \quad \mu_F = 16.084$$

(b) If the mean is now set at 16.084, the contents of an individual bag, X, are normally distributed with $\mu = 16.084$ and $\sigma = .122$. Then

$$P[X < 16.0] = P[Z < \frac{16.0 - 16.084}{.122}] = P[Z < -.689] = .245$$

In the long run, the relative frequency (proportion of underweight bags)

$$\frac{\text{\# bags less than 16.0 oz.}}{\text{total \# bags}}$$

will approach the probability .245.

7.21   (a) We have $E(\bar{X}) = \mu = 31,000$ and $\text{sd}(\bar{X}) = \frac{\sigma}{\sqrt{n}} = \frac{5000}{\sqrt{100}} = 500$.

Since $n = 100$ is large, the central limit theorem ensures that the distribution of $\bar{X}$ is approximately normal with mean and sd as calculated above.

(b) The standardized variable is $Z = \frac{\bar{X} - 31,000}{500}$.

$$P[\bar{X} > 31,500] = P[Z > \frac{31,500 - 31,000}{500}] = P[Z > 1] = .1587.$$

7.23 The population of fry has $\mu = 3.4$ and $\sigma = .8$ so

$$E(\bar{X}) = 3.4 \text{ and } \text{sd}(\bar{X}) = \frac{\sigma}{\sqrt{n}} = \frac{.8}{\sqrt{36}} = .1333$$

and the standardized variable is $Z = \frac{\bar{X} - 3.4}{.1333}$

(a) $P[\bar{X} < 3.2] = P[Z < \frac{3.2 - 3.4}{.1333}] = P[Z < -1.50] = .0668$

(b) Those caught in the net may be the slower less active fish or even the less healthy. Consequently, they may tend to be on the smaller side of the distribution.

7.25 We have $E(\bar{X}) = 34.5$ and $\text{sd}(\bar{X}) = \frac{\sigma}{\sqrt{n}} = \frac{1.3}{\sqrt{6}} = .5307$.

So the standardized variable is $Z = \frac{\bar{X} - 34.5}{.5307}$.

$$P[34.1 < \bar{X} < 35.2] = P[\frac{34.1 - 34.5}{.5307} < Z < \frac{35.2 - 34.5}{.5307}]$$

$$= P[Z < 1.319] - P[Z < -.754] = .906 - .225 = .681$$

7.27 (a) By column, the medians are

```
6   4   4   9   6   6   6
7   4   6   4   4   1   6
4   4   4   5   7   6   5
6   4   7   4   6   7   1
4   6   4   2   5   9   5

4   2   4   7   5   3   8
4   2   2   2   4   7   2
7   5   5   4   5   2   4
6   6   5   4   4   4   3
6   2   4   5   4   4   2

5   3   6   4   1   7
3   5   4   4   5   3
5   8   4   6   6   4
6   4   4   8   3   3
5   5   2   7   5   6
```

(b) & (c) Histograms are given below. The mean has smaller variance.

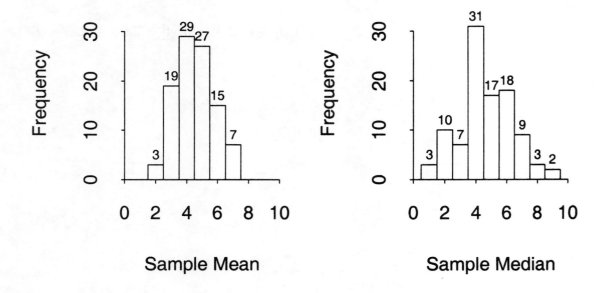

Sample Mean   Sample Median

7.29   (a) and (b)

x	$f(x)$	$xf(x)$	$x^2f(x)$
0	.4	0	0
1	.3	.3	.3
2	.1	.2	.4
3	.2	.6	1.8
Total	1.0	1.1	2.5

Consequently, $E[X] = 1.1$ and s.d. $(X) = \sqrt{2.5 - (1.1)^2} = \sqrt{1.29} = 1.136$.

(c) Let X_1 be the number of complaints on the first day and X_2 be the number of complaints on the second day. The total number of complaints is the sum X_1+X_2. For the outcome (1, 2) the total is $1+2 = 3$ and, since X_1 and X_2 are independent, the probability of the outcome (1, 2) is the product $.3 \times .1 = .03$ and so on. All possible samples (x_1, x_2), the corresponding total values, and the probabilities are:

(x_1, x_2)	$(0,0)$	$(0,1)$	$(0,2)$	$(0,3)$	$(1,0)$	$(1,1)$	$(1,2)$	$(1,3)$
$x_1 + x_2$	0	1	2	3	1	2	3	4
probability	.16	.12	.04	.08	.12	.09	.03	.06

(x_1, x_2)	$(2,0)$	$(2,1)$	$(2,2)$	$(2,3)$	$(3,0)$	$(3,1)$	$(3,2)$	$(3,3)$
$x_1 + x_2$	2	3	4	5	3	4	5	6
probability	.04	.03	.01	.02	.08	.06	.02	.04

$x_1 + x_2$	Probability
0	.16
1	.24
2	.17
3	.22
4	.13
5	.04
6	.04
Total	1.00

(d) Note that, if the number of number of complaints is more than 125, the sample mean will be greater than 125/90. Since the sample size $n = 90$ is large, we approximate the distribution of $\overline{X}$ by a normal distribution with mean $= 1.1$ and standard deviation $= \sigma/\sqrt{n} = 1.136/\sqrt{90}$.

$$P[\overline{X} > \frac{125}{90}] = P[Z > \frac{\frac{125}{90} - 1.1}{1.136/\sqrt{90}}] = P[Z > 2.413] = .0079$$

7.31 (a)

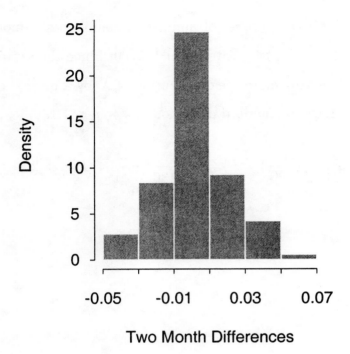

(b) There is more variability in the differences two months apart than in the one month differences. A computer calculation gives the standard deviation .0197 versus .0128 for the one month differences.

7.33   (a) We record here the value of R for each possible sample listed in Exercise 7.32(a).

Values of R for the

Individual samples

$$
\begin{array}{cccc}
0 & 2 & 4 & 6 \\
2 & 0 & 2 & 4 \\
4 & 2 & 0 & 2 \\
6 & 4 & 2 & 0
\end{array}
$$

(b) For the sampling distribution or R we list the distinct values and the corresponding probabilities.

Value of R	Probability
0	4/16
2	6/16
4	4/16
6	2/16
Total	1

7.35 We have $\mu = 60, \sigma = 10$, and $n = 9$.

(a) $E(\bar{X}) = \mu = 60, \quad sd(\bar{X}) = \dfrac{\sigma}{\sqrt{n}} = \dfrac{10}{\sqrt{9}} = \dfrac{10}{3}$

(b) Since the population is normal, $\bar{X}$ has exactly a normal distribution with the mean and sd given in part (a).

(c) The standardized variable is $Z = \dfrac{\bar{X} - 60}{10/3}$

$P[56 < \bar{X} < 64] = P[\dfrac{56 - 60}{10/3} < Z < \dfrac{64 - 60}{10/3}]$

$$= P[-1.2 < Z < 1.2] = .8849 - .1151 = .7698.$$

7.37 We have $\mu = 12.1$, $\sigma = 3.2$, and $n = 9$.

(a) Since the population is normal, the distribution of $\bar{X}$ is exactly normal with mean $= \mu = 12.1$ and sd $= \dfrac{\sigma}{\sqrt{n}} = \dfrac{3.2}{\sqrt{9}} = \dfrac{3.2}{3}$.

(b) The standardized normal variable is $Z = \dfrac{\bar{X} - 12.1}{3.2/3}$.

$$P[\bar{X} < 10] = P[Z < \frac{10 - 12.1}{3.2/3}] = P[Z < -1.97] = .0244.$$

(c) The pebble size X is normally distributed with mean 12.1 and sd 3.2 so $Z = \dfrac{X - 12.1}{3.2}$ is $N(0,1)$.

$$P[X < 10] = P[Z < \frac{10 - 12.1}{3.2}] = P[Z < -.656] = .256.$$

About 26% of the pebbles are of size smaller than 10.

7.39 We have $\mu = 54$, $\sigma = 8$, and $n = 81$.

(a) $E(\bar{X}) = \mu = 54$ and $\mathrm{sd}(\bar{X}) = \dfrac{\sigma}{\sqrt{n}} = \dfrac{8}{\sqrt{81}} = \dfrac{8}{9} = .889$

(b) Since $n = 81$ is large, the central limit theorem asserts that the distribution of $\bar{X}$ is nearly normal with the mean and sd given in part (a).

7.41 (a) Since $n = 45$ is large, the distribution of $\bar{X}$ is nearly normal with

mean $= \mu = 55$, and sd $= \dfrac{\sigma}{\sqrt{n}} = \dfrac{7}{\sqrt{45}} = 1.0435$.

$$P[54 < \bar{X} < 56] = P[\tfrac{54 - 55}{1.0435} < Z < \tfrac{56 - 55}{1.0435}]$$

$$= P[-.958 < Z < .958] = .8310 - .1690 = .6620.$$

(b) For a standard normal variable Z we have

$$P[-1.96 < Z < 1.96] = .95$$

The required interval for $\bar{X}$ is

$$55 \pm 1.96 \ \mathrm{sd}(\bar{X}) = 55 \pm 1.96(1.0435) = 55 \pm 2.05 \quad \text{or} \quad (52.95, 57.05)$$

7.43 (a) and (b)

x	$f(x)$	$xf(x)$	$x^2f(x)$
0	.5	0	0
1	.3	.3	.3
2	.2	.4	.8
Total	1.0	0.7	1.1

Consequently, $E[X] = .7$ and s.d. $(X) = \sqrt{1.1 - (.7)^2} = \sqrt{.61} = .781.$

(c) Let X_1 be the number sold the next day and X_2 be the number sold the day after next. for the return receipt to arrive. The total number sold is the sum $X_1 + X_2$. For the outcome (2, 1) the total is $2 + 1 = 3$ and, since X_1 and X_2 are independent, the probability of the outcome (2, 1) is the product $.2 \times .3 = .06$ and so on. All possible samples (x_1, x_2), the corresponding total values, and the probabilities are:

(x_1, x_2)	$(0,0)$	$(0,1)$	$(0,2)$	$(1,0)$	$(1,1)$	$(1,2)$	$(2,0)$	$(2,1)$	$(2,2)$
$x_1 + x_2$	0	1	2	1	2	3	2	3	4
probability	.25	.15	.10	.15	.09	.06	.10	.06	.04

$x_1 + x_2$	Probability
0	.25
1	.30
2	.29
3	.12
4	.04
Total	1.00

(d) The event that at least 53 kyacks are sold is the same event that $\bar{X} > 53/64$. Since the sample size $n = 64$ is large, the distribution of $\bar{X}$ is approximately normal with mean .7 and standard deviation $= \frac{\sigma}{\sqrt{n}} = \frac{.781}{\sqrt{64}} = .09763$

$$P[\bar{X} > \frac{53}{64}] = P[Z > \frac{\frac{53}{64} - .7}{.09763}] = P[Z > 1.312] = .1948$$

(e) We must have $z-$value 1.645. If k is the required number, then

$$\frac{\frac{k}{64} - .7}{.09763} = 1.645$$

or $k = 64(.7 + 1.645(.09763)) = 55.1$ so 56 kyacks must be ordered.

7.45 (a) Let X be the amount in a single bottle. The distribution of X is normal so the probability is

$$P[X < 299] = P[\frac{X - 302}{2} < \frac{299 - 302}{2}] = P[Z < -1.5] = .0668$$

(b) Since $P[Z > 1.645] = .0500$,

$$\frac{v - 302}{2} = 1.645 \quad \text{or} \quad v = 302 + 2 \times 1.645 = 305.29 \text{ ml}$$

(c) Since the population is normal, $\bar{X}$ has a normal distribution with mean 302 and standard deviation $2/\sqrt{2} = \sqrt{2}$.

$$P[\bar{X} < 299] = P[Z < \frac{299 - 302}{\sqrt{2}}] = P[Z < -2.121] = .017$$

(d) By Part (c), each package has probability .017 of containing less than 299 ml. Let Y be the number out of 2 packages that contain less than 299 ml. Then, Y has a binomial distribution with $n = 2$ and $p = .017$. We are interested in the event $Y = 1$. By the formula for the binomial distribution, the probability is

$$2(.017)(1 - .017) = .0334$$

Chapter 8

DRAWING INFERENCES FROM LARGE SAMPLES

8.1 (a) (i) $S.E. = \frac{\sigma}{\sqrt{n}} = \frac{22}{\sqrt{152}} = 1.784$

      (ii) 95% error margin $= z_{.025}\frac{\sigma}{\sqrt{n}} = 1.96 \times 1.784 = 3.50$

  (b) (i) $S.E. = \frac{8.2}{\sqrt{85}} = .889$

      (ii) 99% error margin $= z_{.005}\frac{\sigma}{\sqrt{n}} = 2.58 \times .889 = 2.29$

  (c) (i) $S.E. = \frac{56}{\sqrt{295}} = 3.260$

      (ii) 92% error margin $= z_{.04}\frac{\sigma}{\sqrt{n}} = 1.75 \times 3.260 = 5.71$

8.3 Point estimate is $\overline{x}$. Estimated standard error is $s/\sqrt{n}$ where

$$s^2 = \frac{\Sigma(x_i - \bar{x})^2}{n - 1}.$$

(a) $\bar{x} = \frac{752}{70} = 10.74$, $s^2 = \frac{235}{69} = 3.406$, $\frac{s}{\sqrt{70}} = .221$

(b) $\bar{x} = \frac{2653}{90} = 29.48$, $s^2 = \frac{546}{89} = 6.135$, $\frac{s}{\sqrt{90}} = .261$

(c) $\bar{x} = \frac{3985}{160} = 24.91$, $s^2 = \frac{745}{159} = 4.686$, $\frac{s}{\sqrt{160}} = .171$

8.5 Estimated mean weakly amount contested $\bar{x} = \$75.43$

Estimated standard error $= \frac{s}{\sqrt{n}} = \frac{24.73}{\sqrt{50}} = 3.497$

90% error margin $= 1.645\frac{s}{\sqrt{n}} = 1.645 \times 3.497 = \$5.75$

8.7 (a) Given that the 95% error margin $= 1.96\frac{s}{\sqrt{n}} = 3.26$,

we obtain the estimated standard error $\frac{s}{\sqrt{n}} = \frac{3.26}{1.96} = 1.663$.

(b) 90% error margin $= 1.645\frac{s}{\sqrt{n}} = 2.74$.

8.9 95% error margin is given by $1.96\dfrac{\sigma}{\sqrt{n}}$

    (a) In order that the 95% error margin be $\frac{1}{8}\sigma$, we must have

$$1.96\frac{\sigma}{\sqrt{n}} = \frac{1}{8}\sigma \text{ or } n = (1.96 \times 8)^2 = 245.9.$$

The required sample size is $n = 246$.

    (b) $1.96\dfrac{\sigma}{\sqrt{n}} = .15\sigma$ or $n = (\frac{1.96}{.15})^2 = 170.7$

The required sample size $n = 171$.

8.11 We have $d = 5.00, \sigma = 25$ and $z_{\alpha/2} = z_{.01} = 2.33$.

We calculate $[\frac{2.33 \times 25}{5.00}]^2 = 135.7$ so $n = 136$.

8.13 Here $d = .5, \sigma = 2.5$ and $z_{\alpha/2} = z_{.05} = 1.645$.

We calculate $[\frac{1.645 \times 2.5}{.5}]^2 = 67.7$ so $n = 68$.

8.15 With $1 - \alpha = .90$ we have $\alpha = .10$ and $z_{\alpha/2} = z_{.05} = 1.645$.

A 90% confidence interval for μ is calculated as

$$\bar{x} \pm 1.645\frac{s}{\sqrt{n}} = 81.3 \pm 1.645\frac{5.8}{\sqrt{63}}$$

$$= 81.3 \pm 1.20 \text{ or } (80.10, 82.50)$$

8.17 95% of 365, or 347 confidence intervals are expected to cover the true means. Before the data are obtained each day, the probability of covering the population mean is .95. By the long run frequency interpretation of probability, approximately 95 % will cover.

8.19 For large n, a 95% confidence interval for μ is given by $\overline{X} \pm z_{.025}S/\sqrt{n}$. Using $z_{.025} = 1.96, n = 35$ and the summary statistics $\overline{x} = 30.2$ grams, $s = 3.8$ grams, the 95% confidence interval for μ is found to be $30.2 \pm 1.96\dfrac{3.8}{\sqrt{35}} = 30.2 \pm 1.26$ or $(28.94, 31.46)$ grams.

8.21 For 99% confidence interval we need $z_{\alpha/2} = z_{.005} = 2.58$. With $n = 120$ and the summary statistics $\overline{x} = 18.3$ days, $s = 5.2$ days, a 99% confidence interval for the true mean survival time is found to be

$$18.3 \pm 2.58 \frac{5.2}{\sqrt{120}} = 18.3 \pm 1.2 \text{ or } (17.1, 19.5) \text{ days.}$$

8.23 We have $n = 70, \overline{x} = 493$ and $s = 72$. For $1-\alpha = .95$ we have $z_{\alpha/2} = z_{.025} = 1.96$. A 95% confidence interval for the population mean math score is then

$$493 \pm 1.96 \frac{72}{\sqrt{70}} = 493 \pm 16.9 \text{ or } (476.1, 509.9).$$

8.25 We have $n = 140, \overline{x} = 8.6$ miles and $s = 4.3$ miles. For $1 - \alpha = .90$ we have $z_{\alpha/2} = z_{.05} = 1.645$. A 90% confidence interval for the population mean commuting distance is then

$$8.6 \pm 1.645 \frac{4.3}{\sqrt{140}} = 8.6 \pm .60 \text{ or } (8.00, 9.20) \text{ miles.}$$

8.27 We have $n = 50, \overline{x} = 75.43$ dollars and $s = 24.73$ dollars. For $1 - \alpha = .95$, we have $z_{\alpha/2} = z_{.025} = 1.96$. A 95% confidence interval for the population mean amount contested is then

$$75.43 \pm 1.96 \frac{24.73}{\sqrt{50}} = 75.43 \pm 6.85 \text{ or } (68.58, 82.28) \text{ dollars.}$$

8.29 For 95% confidence we require $\alpha = .05$ so $z_{\alpha/2} = z_{.025} = 1.96$. The required confidence interval is calculated as

$$\bar{x} \pm 1.96\frac{s}{\sqrt{n}} = .0011 \pm 1.96\frac{.0128}{\sqrt{179}}$$

$$= .0011 \pm .0019 \text{ or } (-.0008, .0030).$$

8.31 We have $n = 40, \bar{x} = 1.715$ centimeters and $s = .475$ cm.

For $1 - \alpha = .99$ we have $\alpha = .01$ and $z_{\alpha/2} = z_{.005} = 2.58$. A 99% confidence interval for μ is then

$$1.715 \pm 2.58\frac{.475}{\sqrt{40}} = 1.715 \pm .194 \text{ or } (1.521, 1.909) \text{ centimeters.}$$

8.33 We have $\bar{x} = 126.7$ and $1.96\frac{s}{\sqrt{n}} = 5.8$

(a) For large n, a 95% confidence interval for μ is calculated as

$$\bar{x} \pm 1.96\frac{s}{\sqrt{n}} = 126.7 \pm 5.8 \text{ or } (120.9, 132.5)$$

(b) A 90% confidence interval for μ is given by $\overline{X} \pm 1.645\frac{S}{\sqrt{n}}$.

Because $1.96\frac{s}{\sqrt{n}} = 5.8$, we have $\frac{s}{\sqrt{n}} = \frac{5.8}{1.96}$ so the required confidence interval becomes

$$126.7 \pm 1.645 \frac{5.8}{1.96} = 126.7 \pm 4.9 \text{ or } (121.8, 131.6).$$

8.35 For large n, a $100(1-\alpha)\%$ confidence interval for μ is given by $(\overline{X} - z_{\alpha/2} \frac{S}{\sqrt{n}}, \overline{X} + z_{\alpha/2} \frac{S}{\sqrt{n}})$. For this interval to be the same as $(\overline{X} - .2S, \overline{X} + .2S)$ when $n = 81$, we must have $z_{\alpha/2} \frac{S}{\sqrt{81}} = .2S$ or $z_{\alpha/2} = .2\sqrt{81} = 1.80$.

From the standard normal table we find $P[Z > 1.80] = .0359$ so $\frac{\alpha}{2} = .0359, \alpha = .0718$ or $1 - \alpha = .9282$.

The level of confidence is .9282 or about 93%.

8.37 The alternative hypothesis H_1 is the assertion that is to be established; its opposite is the null hypothesis H_0.

(a) Let μ denote the population mean time, in days, to pay a claim. The hypotheses are

$$H_0 : \mu = 14, \quad H_1 : \mu < 14$$

(b) Let μ denote the population mean amount spent (in dollars). The hypotheses are

$$H_0 : \mu = 2.50, \quad H_1 : \mu > 2.50$$

(c) Let $\mu =$ denote the population mean hospital bill (in dollars). The hypotheses are

$$H_0 : \mu = 3000, \quad H_1 : \mu < 3000$$

(d) Let $\mu =$ denote the population mean time between purchases (in days). The hypotheses are

$$H_0 : \mu = 60, \quad H_1 : \mu \neq 60.$$

8.39 Retaining H_0 is a correct decision if H_0 is true. It is a wrong decision if H_1 is true, and in that case a type II error is made.

(a) Correct decision if $\mu = 14$. Wrong decision if $\mu < 14$, Type II error.

(b) Correct decision if $\mu = 2.50$. Wrong if $\mu > 2.50$, Type II error.

(c) Correct decision if $\mu = 3000$. Wrong if $\mu < 3000$, Type II error.

(d) Correct decision if $\mu = 60$. Wrong if $\mu \neq 60$, Type II error.

8.41 (b) (i) $H_0 : \mu = .15,\ H_1 : \mu < 1.5$

(ii) $Z = \dfrac{\overline{X} - .15}{.085/\sqrt{125}}$,   (iii) $R : Z \leq -1.96$

(c) (i) $H_0 : \mu = 80,\quad H_1 : \mu \neq 80$

(ii) $Z = \dfrac{\overline{X} - 80}{8.6/\sqrt{38}}$,   (iii) $R : |Z| \geq 2.58$

(d) (i) $H_0 : \mu = 0,\quad H_1 : \mu \neq 0$

(ii) $Z = \dfrac{\overline{X}}{1.23/\sqrt{40}}$,   (iii) $R : |Z| \geq 1.88$.

8.43 Because the claim is that $\mu > 30$, we formulate the hypotheses

$$H_0 : \mu = 30, \quad H_1 : \mu > 30$$

With $n = 70$ and $\sigma = 5.6$, the test statistic (in standardized form) is

$$Z = \frac{\overline{X} - 30}{5.6/\sqrt{70}}$$

(a) $\overline{X} \geq 31.31$ is equivalent to

$$Z \geq \frac{31.31 - 30}{5.6/\sqrt{70}} \text{ or } Z \geq 1.96$$

Since $P[Z \geq 1.96] = .025$, the level of significance is $\alpha = .025$

(b) Because $P[Z \geq 1.645] = .05$, the rejection region $R : Z \geq 1.645$ has the level of significance $\alpha = .05$.

Using the relation $Z = \dfrac{\overline{X} - 30}{5.6/\sqrt{70}}$

We note that $Z \geq 1.645$ is equivalent to

$$\overline{X} \geq 30 + 1.645\frac{5.6}{\sqrt{70}}$$

or $\overline{X} \geq 31.10$

Therefore $c = 31.10$.

8.45 The test statistic is the same, but the rejection region is now two-sided. With $\alpha = .05$, we have $z_{\alpha/2} = z_{.025} = 1.96$ and the two-sided rejection region R : $|Z| \geq 1.96$. The observed value $|z| = 1.88$ is not in this rejection region. Hence, H_0 is not rejected at $\alpha = .05$. The $P-$ value is $2P[Z \leq -1.88] = 2(.0301) = .0602$.

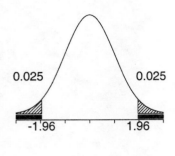

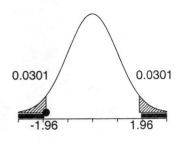

(a) Rejection Region   (b) P-value for Problem 8.45

8.47 Because H_1 is left-sided and the observed $z = -1.81$, the significance probability

or

$$\text{P–value} = P[Z \le -2.41] = .0080$$

This means that H_0 would be rejected with α as small as .008 which is less than 1%. A small P–value signifies a strong rejection of H_0.

8.49 Because the claim is that $\mu > 3.5$, we formulate the hypotheses

$$H_0 : \ \mu = 3.5 \text{ vs. } H_1 : \ \mu > 3.5$$

The sample size is $n = 40$. The test statistic is

$$Z = \frac{\overline{X} - 3.5}{S/\sqrt{40}}$$

and the rejection region is of the form $R : \ Z \geq c$ because H_1 is right-sided. Since $z_{.10} = 1.28$, the rejection region with $\alpha = .10$ is set as $R : \ Z \geq 1.28$.

With the observed values $\overline{x} = 3.8$ and $s = 1.2$, we calculate

$$z = \frac{3.8 - 3.5}{1.2/\sqrt{40}} = 1.58$$

Since the observed $z = 1.58$ is in R, H_0 is rejected. We conclude that the claim is substantiated at $\alpha = .10$. The $P-$value is $P[Z \geq 1.58] = .0571$.

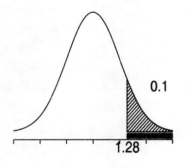
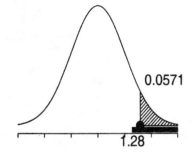
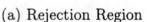

(a) Rejection Region   (b) P-value for Problem 8.49

8.51 The sample size is $n = 36$ and the test statistic is

$$Z = \frac{\overline{X} - \mu_0}{S/\sqrt{36}}$$

(a) Here $\mu_0 = 74$, the test statistic is

$$Z = \frac{\overline{X} - 74}{S/\sqrt{36}}$$

and the rejection region consists of large values of Z because H_1 is right-sided.

Using $\overline{x} = 80.4$ and $s = 16.2$, the observed value of the test statistic is

$$z = \frac{80.4 - 74}{16.2/\sqrt{36}} = 2.37$$

P–value $= P[Z \geq 2.37] = .0089$. This extremely small P–value signifies a strong rejection of H_0.

(b) Here $\mu_0 = 85$, the test statistic is

$$Z = \frac{\overline{X} - 85}{S/\sqrt{36}}$$

and small values of Z should form the rejection region because H_1 is left-sided. The observed value of the test statistic is

$$z = \frac{80.4 - 85}{16.2/\sqrt{36}} = -1.70$$

P–value $= P[Z \leq -1.70] = .0446$

The null hypothesis would be rejected for α as small as .0446 which is about 4%. There is a fairly strong evidence for rejection of H_0.

(c) Here $\mu_0 = 76$, the test statistic is

$$Z = \frac{\overline{X} - 76}{S/\sqrt{36}}$$

and the rejection region consists of large values of $|Z|$ because H_1 is two-sided. The observed test statistic is

$$z = \frac{80.4 - 76}{16.2/\sqrt{36}} = 1.63$$

P–value $= P[|Z| \geq 1.63] = 2P[Z \leq -1.63] = 2 \times .0516 = .1032.$

The smallest α at which H_0 can be rejected is .1032. Since this is not very small, the support for H_1 is not strong.

8.53 Let μ denote the population mean hold time (in minutes). Because the claim is that $\mu > 3.0$, we formulate

$$H_0 : \mu = 3.0 \text{ vs. } H_1 : \mu > 3.0$$

The sample size is $n = 75$. The test statistic is

$$Z = \frac{\overline{X} - 3.0}{S/\sqrt{75}}$$

The rejection region should consist of large values of Z because H_1 is right-sided. Since $z_{.05} = 1.645$, the rejection region with $\alpha = .05$ is set as $R : Z \geq 1.645$.

With the observed values $\overline{x} = 3.4$ and $s = 2.4$, we calculate

$$z = \frac{3.4 - 3.0}{2.4/\sqrt{75}} = 1.44$$

The value $z = 1.44$ is not in R so H_0 is not rejected. The claim is not substantiated at $\alpha = .05$. The P–value $= P[Z \geq 1.44] = .0749$. Since this is not small, support for the claim is weak.

8.55 Let μ denote the true mean BOD. Because the intent is to establish that μ is different from 3000, we formulate

$$H_0 : \ \mu = 3000 \text{ vs. } H_1 : \ \mu \neq 3000$$

Letting $\overline{X}$ and S denote the sample mean and standard deviation for $n = 43$ measurements, the test statistic is

$$Z = \frac{\overline{X} - 3000}{S/\sqrt{43}}$$

The rejection region should be of the form $R : \ |Z| \geq z_{\alpha/2}$ because H_1 is two-sided. With $\alpha = .05$ we have $\alpha/2 = .025$ and $z_{.025} = 1.96$ so we set $R : |Z| \geq 1.96$.

With the observed values $\overline{x} = 3246$ and $s = 757$ we calculate

$$z = \frac{3246 - 3000}{757/\sqrt{43}} = 2.13$$

Since the value $z = 2.13$ lies in R, H_0 is rejected at $\alpha = .05$. This means that there is strong evidence that the BOD is significantly off the target.

8.57 (a) Point estimate $\hat{p} = \frac{28}{50} = .56$

Estimated $S.E. = \sqrt{\frac{\hat{p}\hat{q}}{n}} = \sqrt{\frac{.56 \times .44}{50}} = .070$

95% error margin = 1.96 (estimated $S.E.$)

$$= 1.96 \times \sqrt{\frac{.56 \times .44}{50}} = .138$$

(b) Point estimate $\hat{p} = \frac{75}{460} = .163$

95% error margin $= 1.96\sqrt{\frac{.163 \times .837}{460}} = .034$

(c) Point estimate $\hat{p} = \frac{2001}{2500} = .800$

95% error margin $= 1.96\sqrt{\frac{.800 \times .200}{2500}} = .016$

8.59 A 90% confidence interval for p is given by

$$\hat{p} \pm 1.645\sqrt{\hat{p}\hat{q}/n}$$

The results are

(a) $.56 \pm 1.645\sqrt{\frac{.56 \times .44}{50}} = .56 \pm .12$ or $(.44, .68)$

(b) $.163 \pm 1.645\sqrt{\frac{.163 \times .837}{460}} = .163 \pm .028$ or $(.13, .19)$

(c) $.800 \pm 1.645\sqrt{\frac{.800 \times .200}{2500}} = .800 \pm .013$ or $(.79, .81)$

8.61 (a) The estimate of the market share is $\hat{p} = 120/325 = .369$

(b) Estimated $S.E. = \sqrt{\frac{\hat{p}\hat{q}}{n}} = \sqrt{\frac{.369 \times .631}{325}} = .027.$

8.63 Here $n = 128$ and $\hat{p} = 72/128 = .5625$. For $p =$ population proportion of infected rodents, a 95% confidence interval is

$$\hat{p} \pm 1.96\sqrt{\hat{p}\hat{q}/n} = .5625 \pm 1.96\sqrt{(.5625 \times .4375)/128}$$

$$= .5625 \pm .0860 \text{ or } (.48, .65).$$

8.65 Let $p =$ the population proportion of all possible policies that would have at least one claim in a year.

Point estimate $\hat{p} = \dfrac{2073}{12299} = .169.$

90% error margin $= z_{.05}\sqrt{\dfrac{\hat{p}\hat{q}}{n}} = 1.645\sqrt{\dfrac{.169 \times .831}{12299}} = .006$

8.67 (a) Denote $p_M =$ population proportion of ERS calls involving serious mechanical problems. Since 849 calls involved serious mechanical problems out of $n = 2927$ ERS calls, the estimate of p_M is $\hat{p}_M = \dfrac{849}{2927} = .290.$

95% error margin $= 1.96\sqrt{\dfrac{.29 \times .71}{2927}} = .016.$

(b) Denote p_S =population proportion of ERS calls involving starting problems. Since there were 1499 calls related to starting problems out of $n = 2927$ calls, we obtain $\hat{p}_S = \frac{1499}{2927} = .512$, and $\hat{q}_S = 1 - .512 = .488$.

98% confidence interval for p_S is then calculated as

$$\hat{p}_S \pm 2.33 \sqrt{\frac{\hat{p}_S \hat{q}_S}{n}}$$

$$= .512 \pm 2.33 \sqrt{\frac{.512 \times .488}{2927}}$$

$$= .512 \pm .022 \text{ or } (.49, .53).$$

8.69 The tolerable error is $d = .03$. Also $1 - \alpha = .90$ gives $\alpha/2 = .05$ and $z_{\alpha/2} = 1.645$.

(a) With $p* = .2$, the required sample size is

$$n = (.2 \times .8)[\frac{1.645}{.03}]^2 = 481.1, \text{ that is, } n = 482.$$

(b) Without any knowledge of p, the conservative solution is

$$n = \frac{1}{4}[\frac{1.645}{.03}]^2 = 751.7.$$

A sample of size $n = 752$ would suffice.

8.71 (a) We have $\hat{p} = .32$, and the 90% error margin

$$1.645\sqrt{\frac{\hat{p}\hat{q}}{n}} = .08 \quad \text{so} \quad \sqrt{\frac{\hat{p}\hat{q}}{n}} = \frac{.08}{1.645}$$

A 95% confidence interval for p is

$$\hat{p} \pm 1.96\sqrt{\frac{\hat{p}\hat{q}}{n}} = .32 \pm 1.96[\frac{.08}{1.645}]$$

$$= .32 \pm .10 \text{ or } (.22, .42)$$

(b) Again referring to the 90% error margin, and the given numerical value $\hat{p} = .32$, we have

$$.08 = 1.645\sqrt{\frac{\hat{p}\hat{q}}{n}} = 1.645\sqrt{\frac{.32 \times .68}{n}}$$

so $n = [\frac{1.645}{.08}]^2(.32 \times .68) = 92.0$

The study was based on a sample of size $n = 92$.

8.73 (a) Let $p = $ the proportion of smokers who would quit on the first attempt.

$H_0 : p = .40$ and $H_1 : p < .40$.

(b) Let p = the proportion of drivers that use the lane illegally. $H_0 : p = .25$ and $H_1 : p > .25$.

(c) Let p = the proportion of patients that would wait over half and hour. $H_0 : p = .20$ and $H_1 : p < .20$.

8.75 (a) $z = \frac{.233 - .32}{.0426} = -2.04$. Because -2.04 lies in $R : Z \leq -1.645, H_0$ is rejected at $\alpha = .05$.

(b) $z = \frac{.817 - .75}{.0287} = 2.33$. Because 2.33 is in $R : Z \geq 2.05, H_0$ is rejected at $\alpha = .02$

(c) $z = \frac{.709 - .60}{.0558} = 1.95$. Because 1.95 is not in $R : |Z| \geq 2.33, H_0$ is not rejected at $\alpha = .02$

(d) $z = \frac{.387 - .56}{.0535} = -3.23$. Because -3.23 lies in $R : Z \leq -1.28, H_0$ is rejected at $\alpha = .10$.

8.77 (a) Test statistic: $Z = \dfrac{\hat{p} - .3}{\sqrt{(.3 \times .7)/n}}$

Since $z_{.05} = 1.645$ and H_1 is right-sided, the rejection region is $R : Z \geq 1.645$.

(b) The observed sample proportion is $\hat{p} = 19/48$, and the test statistic has the value

$$z = \dfrac{(19/48) - .3}{\sqrt{(.3 \times .7)/48}} = 1.45, \text{ not in } R.$$

With $\alpha = .05$, we do not reject H_0.

P–value $= P[Z \geq 1.45] = .0735 \approx .07$.

Therefore, α must be taken larger than .07 in order to reject H_0.

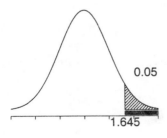

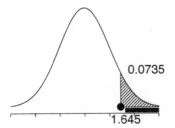

(a) Rejection Region   (b) P-value for Problem 8.77

8.79 Let p denote the population proportion of ERS calls involving flat tires or lockouts. Because we want to establish that p is smaller than .19, we formulate the hypotheses

$$H_0 : \ p = .19 \text{ vs. } H_1 : \ p < .19$$

Denoting $\hat{p} =$ the sample proportion based on $n = 2927$, the test statistic is

$$Z = \frac{\hat{p} - .19}{\sqrt{(.19 \times .81)/2927}} = \frac{\hat{p} - .19}{.00725}$$

Because H_1 is left-sided, the rejection should be of the form $R : \ Z \leq -z_\alpha$. With $\alpha = .05$ the rejection region is $R : \ Z \leq -1.645$.

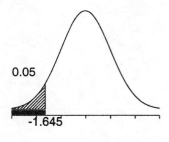

Using the sample proportion $\hat{p} = \frac{498}{2927} = .170$ we calculate

$$z = \frac{.170 - .19}{.00725} = -2.76$$

The observed $z = -2.76$ lies in R so H_0 is rejected at $\alpha = .05$. We conclude that $\hat{p}$ is significantly smaller than $.19$.

8.81 Let p denote population proportion of persons who would rate the services excellent. We are to test

$$H_0 : \; p = .46 \text{ vs. } H_1 : \; p \neq .46$$

Denoting $\hat{p} =$ the sample proportion based on $n = 505$, the test statistic is

$$Z = \frac{\hat{p} - .46}{\sqrt{(.46 \times .54)/505}} = \frac{\hat{p} - .46}{.0222}$$

Because H_1 is two-sided, the rejection region should be of the form $R : \; |Z| \geq z_{\alpha/2}$. Taking $\alpha = .10$, the rejection region is as $R : \; Z \leq -1.645$ or $Z \geq 1.645$.

Using the sample proportion $\hat{p} = \frac{258}{505} = .5109$, we calculate

$$z = \frac{.5109 - .46}{.0222} = 2.29$$

The observed $z = 2.29$ lies in R so H_0 is rejected at $\alpha = .10$.

We also calculate

$$\text{P–value} = P[Z \le -2.29] + P[Z \ge 2.29] = .0220$$

This low P–value signifies a strong support of H_1.

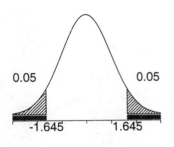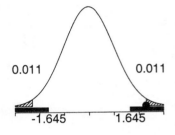

(a) Rejection Region   (b) P-value for Problem 8.81

8.83   (a) We have $n = 505$ and $\hat{p} = \frac{258}{505} = .5109$. A 90% confidence interval for p is

calculated as

$$\hat{p} \pm z_{.05}\sqrt{\frac{\hat{p}\hat{q}}{n}} = .5109 \pm 1.645\sqrt{\frac{.5109 \times .4881}{505}}$$

$$= .5109 \pm .0366 \text{ or } (.4743, .5475)$$

which we present as $(.47, .55)$.

(b) The unknown total number of customers who would rate the service excellent is $8200p$. The lower endpoint of the 90 % confidence interval for the
total is then $8200(.4743) = 3889$ and the upper limit is $8200(.5475) = 4490$.

8.85 We have $n = 42$ and $\hat{p} = \frac{30}{42} = .7143$. A 95% confidence interval for p is calculated as

$$\hat{p} \pm z_{.025}\sqrt{\frac{\hat{p}\hat{q}}{n}} = .7143 \pm 1.96\sqrt{\frac{.7143 \times .2857}{42}}$$

$$= .7143 \pm .1366 \text{ or } (.58, .85), \text{ that is } (58\%, 85\%).$$

8.87 Point estimate is $\overline{x} = \frac{\Sigma x_i}{n}$. Estimated standard error is $s/\sqrt{n}$ where $s^2 = \Sigma(x_i - \overline{x})^2/(n-1)$.

(a) $\bar{x} = \frac{752}{80} = 9.40$ $s^2 = \frac{345}{79} = 4.367$ $S.E. = \frac{s}{\sqrt{n}} = .234$

(b) $\bar{x} = \frac{1290}{169} = 7.63$ $s^2 = \frac{842}{168} = 5.012$ $S.E. = \frac{s}{\sqrt{n}} = .172$

8.89 The standard error of $\overline{X}$ is $\frac{\sigma}{\sqrt{n}}$ where n is the sample size.

(a) Since $\frac{1}{2}\frac{\sigma}{\sqrt{n}} = \frac{\sigma}{\sqrt{4n}}$, we need a sample of size $4n$.

Therefore, we must increase sample size by a factor of 4.

(b) Since $\frac{1}{4}\frac{\sigma}{\sqrt{n}} = \frac{\sigma}{\sqrt{16n}}$, the sample size must be increased by a factor of 16.

8.91 Since $[\frac{1.96 \times 80}{10}]^2 = 245.8$, the required sample size is 246.

8.93 A $100(1-\alpha)\%$ confidence interval for μ is given by $\overline{X} \pm z_{\alpha/2}\frac{S}{\sqrt{n}}$.

(a) $\bar{x} = \frac{752}{60} = 12.53$, $s^2 = \frac{426}{59} = 7.220$, $\frac{s}{\sqrt{n}} = .347$.

With $1 - \alpha = .95$ we have $\alpha/2 = .025$ and $z_{.025} = 1.96$.

The required confidence interval is calculated as

$$12.53 \pm 1.96(.347) = 12.53 \pm .68 \text{ or } (11.85, 13.21).$$

(b) $\bar{x} = \frac{2562}{150} = 17.08, \quad s^2 = \frac{3722}{149} = 24.98, \quad \frac{s}{\sqrt{n}} = .408.$

With $1 - \alpha = .90$ we have $\alpha/2 = .05$ and $z_{.05} = 1.645$.

The required confidence interval is calculated as

$$17.08 \pm 1.645(.408) = 17.08 \pm .67 \text{ or } (16.41, 17.75).$$

8.95 (a) Correct.

(b) We cannot tell whether this particular interval covers the true mean. It is a single realization of the random interval $\overline{X} \pm 1.645 \frac{S}{\sqrt{n}}$. In repeated sampling, about 90% of the intervals, computed from this formula, will cover μ.

8.97 The alternative hypothesis H_1 is the assertion that is to be established; its opposite is the null hypothesis H_0.

(a) μ = population mean mileage

$H_0 : \mu = 50, \qquad H_1 : \mu < 50$

(b) μ = mean number of pages per transmission

$H_0 : \mu = 3.4, \qquad H_1 : \mu > 3.4$

(c) p = probability of success with the method

$H_0 : p = .5, \qquad H_1 : p > .5$

(d) μ = mean fill

$H_0 : \mu = 16, \qquad H_1 : \mu \neq 16$

(e) μ = mean percent fat content

$H_0 : \mu = 4, \qquad H_1 : \mu > 4$

8.99 Because the claim is that $\mu \neq 32$, we formulate

$$H_0 : \mu = 32 \text{ vs. } H_1 : \mu \neq 32$$

With $n = 100$ and $\sigma = 10.6$, the test statistic (in standardized form) is

$$Z = \frac{\overline{X} - 32}{10.6/\sqrt{100}} = \frac{\overline{X} - 32}{1.06}$$

(a) $|\overline{X} - 32| \geq 2.47$ means that

$$|Z| \geq \frac{2.47}{1.06}, \text{ that is, } |Z| \geq 2.33$$

Since $P[|Z| \geq 2.33] = .02$, we have $\alpha = .02$

(b) Because $P[|Z| \geq 1.96] = .05$, the rejection region $R : |Z| \geq 1.96$ has $\alpha = .05$. Now, $|Z| \geq 1.96$ means that $|\overline{X} - 32| \geq 1.96 \times 1.06$. Therefore,

$$c = 1.96 \times 1.06 = 2.0776.$$

8.101 (a) Test statistic: $Z = \dfrac{\overline{X} - 20}{S/\sqrt{n}}$. Since H_1 is two-sided and $z_{\alpha/2} = z_{.025} = 1.96$, the rejection region is $R : |Z| \geq 1.96$.

(b) Test statistic: $Z = \dfrac{\overline{X} - 30}{S/\sqrt{n}}$. Since H_1 is left-sided, and $z_{.025} = 1.96$, the rejection region is $R : Z \leq -1.96$.

8.103 (a) $E[X] = 0 \times .5 \ + \ 1 \times .3 \ + \ 2 \times .2 = .7$

(b) Since $E[X^2] = 0^2 \times .5 + 1^2 \times .3 + 2^2 \times .2 = 1.1$, $\mathrm{var}(X) = E[X^2] - (E[X])^2 = $
1.1 $- \ (.7)^2 = .61$ so $\sigma = \sqrt{61} = .781$.

(c) We formulate $H_0: \quad \mu = .7$ versus $H_1: \quad \mu > .7$. The test statistic is
$Z = \dfrac{\overline{X} - .7}{S/\sqrt{n}}$ where $n = 64$ and we will use that standard deviation from
the new sample because the distribution of sales may have changed. With
$\alpha = .05$, we have $z_{.05} = 1.645$ so the rejection region is $R: \ Z \geq 1.645$.

The observed value of Z is $z = \dfrac{.84 - .7}{.4/\sqrt{64}} = 2.80$. Since this is lies in R, H_0
is rejected at $\alpha = .05$. The P–value $= P[Z \geq 2.80] = .0026$ so the evidence
in favor of increased mean sales is very strong.

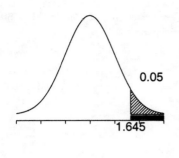

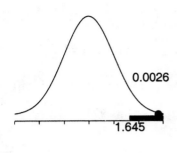

(a) Rejection Region   (b) P-value for Problem 8.103

8.105   (a) Point estimate of proportion unemployed $\hat{p} = \frac{175}{2000} = .0875$

     (b) The 95% error margin uses $z_{.025} = 1.96$ and is given by

$$1.96 \sqrt{\frac{\hat{p}\hat{q}}{n}} = 1.96 \sqrt{\frac{.0875 \times .9125}{2000}} = .0124$$

8.107   (a) The test statistic is $Z = \dfrac{\hat{p} - .25}{\sqrt{\dfrac{.25 \times .75}{n}}}$ where $\hat{p}$ is the sample proportion.
With $\alpha = .05$, the rejection region is $R: Z \geq z_{.05} = 1.645$.

(b) The observed sample proportion is $\hat{p} = \dfrac{65}{190} = .342$, and the test statistic has the value

$$z = \frac{.342 - .25}{\sqrt{(.25 \times .75)/190}} = 2.93$$

which falls in the rejection region set in (a).   Hence H_0 is rejected at $\alpha = .05$. The $P-$value is $P[Z \geq 2.93] = .0017$ so the evidence in favor of the alternative is very strong.

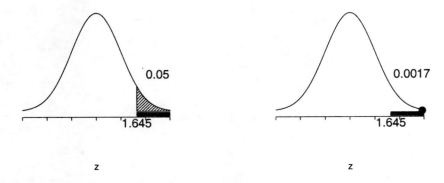

(a) Rejection Region   (b) P-value for Problem 8.107

8.109 Because the claim is that p is different from .3, we formulate

$$H_0 : \ p = .3 \ \text{vs.} \ H_1 : \ p \neq .3$$

Denoting $\hat{p} = $ the sample proportion based on a sample of size $n = 250$, the test statistic is

$$Z = \frac{\hat{p} - .3}{\sqrt{(.3 \times .7)/250}} = \frac{\hat{p} - .3}{.029}$$

(a) $|\hat{p} - .3| \geq .06$ means that $|Z| \geq \frac{.06}{.029}$ or $|Z| \geq 2.07$.

Since $P[|Z| \geq 2.07] = 2P[Z \leq -2.07] = 2 \times .0192 = .0384$, we have $\alpha = .0384$.

(b) Because $P[|Z| \geq 1.645] = .10$, the rejection region with $\alpha = .10$ is given by $R : \ |Z| \geq 1.645$.

Now $|Z| \geq 1.645$ means that

$$|\hat{p} - .3| \geq 1.645(.029)$$

or $|\hat{p} - .3| \geq .048$

Therefore, $c = .048$.

8.111 (a) Let p = population proportion in favor of almond scent. We are to test
$H_0 : p = .5$ vs $H_1 : p \neq .5$. The test statistic is

$$Z = \frac{\hat{p} - .5}{\sqrt{(.5 \times .5)/250}} = \frac{\hat{p} - .5}{.0316}$$

With $\alpha = .05$, the rejection region is $R : |Z| \geq 1.96$. Since 105 out of 250
consumers preferred the almond scent, we have $\hat{p} = 105/250 = .42$. The
observed value of Z,

$$z = \frac{.42 - .5}{.0316} = -2.53$$

lies in the rejection region so H_0 is rejected at $\alpha = .05$. We conclude that
the popularity of the two scents is significantly different.

(b) A 95% confidence interval for p is

$$\hat{p} \pm 1.96\sqrt{\hat{p}\hat{q}/n} = .42 \pm 1.96\sqrt{(.42 \times .58)/250}$$

$$= .42 \pm .06 \text{ or } (.36, .48).$$

8.113 (a) Let p = population proportion of plants that are of the dwarf variety. We want to test $H_0 : p = .8$ vs. $H_1 : p \neq .8$.

The test statistic is

$$Z = \frac{\hat{p} - .8}{\sqrt{(.8 \times .2)/200}} = \frac{\hat{p} - .8}{.0283}$$

and, with $\alpha = .05$, the rejection region is $R : |Z| \geq 1.96$. From the sample data we find $\hat{p} = 136/200 = .68$ and

$$z = \frac{.68 - .8}{.0283} = -4.24.$$

Since $z = -4.24$ lies in the rejection region, H_0 is rejected at $\alpha = .05$. The significance probability

$$P[|Z| \geq 4.24] \quad < \quad .0001$$

is extremely small. A contradiction of the genetic model is strongly indicated.

(b) A 95% confidence interval for p is

$$\hat{p} \pm 1.96\sqrt{\hat{p}\hat{q}/n} = .68 \pm 1.96\sqrt{(.68 \times .32)/200}$$

$$= .68 \pm .06 \text{ or } (.62, .74).$$

*8.115 (a) Under the alternative $\mu_1 = 10.5, \overline{X}$ is normally distributed with mean $= 10.5$ and $sd = \frac{\sigma}{\sqrt{n}} = .25$ so $Z = \frac{\overline{X} - 10.5}{.25}$ is $N(0, 1)$. Therefore, power at $\mu_1 = 10.5$ is

$$P[\overline{X} \geq 10.49 \text{ when } \mu_1 = 10.5] = P[Z \geq \frac{10.49 - 10.5}{.25}]$$

$$= P[Z \geq -.04] = .5160.$$

(b) For $\mu_1 = 10.8, Z = \frac{\overline{X} - 10.8}{.25}$ is $N(0, 1)$ so

$$\text{Power} \quad = P[\overline{X} \geq 10.49 \text{ when } \mu_1 = 10.8]$$

$$= P[Z \geq \frac{10.49 - 10.8}{.25}]$$

$$= P[Z \geq -1.24] = .8925.$$

8.117 The MINITAB output is given below.

```
            N     MEAN   MEDIAN     STDEV   SEMEAN
C1         40   1.7150   1.6000    0.4748   0.0751

          MIN      MAX       Q1       Q3
C1     0.8000   3.1000   1.5000   2.0000
```

(a) From the computer printout we obtain $\bar{x} = 1.715$ and $s = .4748$. The 97% confidence interval for μ is calculated as

$$\bar{x} \pm 2.17s/\sqrt{n} = 1.715 \pm (2.17)(.4748)/\sqrt{40} \quad \text{or} \quad (1.55, \ 1.88)$$

(b) The test statistic is

$$Z = \frac{\bar{X} - \mu_0}{S/\sqrt{n}} = \frac{\bar{X} - 1.9}{S/\sqrt{40}}$$

The alternative is two-sided so the rejection region is $R : |Z| \geq 2.17$ for $\alpha = .03$. Since

$$Z = \frac{\bar{X} - 1.9}{S/\sqrt{40}} = \frac{1.715 - 1.9}{.4748/\sqrt{40}} = -2.464$$

we reject $H_0 : \mu = 1.9$.

Chapter 9

SMALL–SAMPLE INFERENCES FOR NORMAL POPULATIONS

9.1   (a) $t_{.05} = 2.015$ (b) $-t_{.025} = -2.101$

        (c) $-t_{.01} = -2.718$ (d) $t_{.10} = 1.337$

9.3   (a) 90th percentile of $t = t_{.10} = 1.363$

        (b) 99th percentile $= t_{.01} = 4.541$

        (c) 5th percentile $= -t_{.05} = -1.706$

        (d) upper quartile $= t_{.25} = .688$

           lower quartile $= -t_{.25} = -.688$.

9.5 (a) Since the area to the right of b is $1 - .95 = .05$, b is the upper .05 point of the t distribution with d.f. $= 6$. From the t–table we find $b = t_{.05} = 1.943$.

(b) Since $P[T > b] = .025$, we read $t_{.025}$ with d.f. $= 15$ and find $b = t_{.025} = 2.131$.

(c) $b = t_{.01} = 2.896$.

(d) Since $P[T \leq b] = .01$, b is the lower .01 point: $b = -t_{.01} = -2.718$.

9.7 (b) In the t–table with d.f. $= 16$, we look for the percentage points that are close to the given number 1.9 . We find $t_{.05} = 1.746$ and $t_{.025} = 2.120$. Because 1.9 lies between $t_{.025}$ and $t_{.05}$, the probability

$P[T > 1.9]$ must be between .025 and .05 .

(c) The number 1.5 is between $t_{.10} = 1.363$ and $t_{.05} = 1.796$ and $P[T < -1.5] = P[T > 1.5]$. Therefore,

$P[T < -1.5]$ must be between .05 and .10 .

(d) The number 1.9 is between $t_{.05} = 1.812$ and $t_{.025} = 2.228$, and $P[|T| > 1.9] = 2P[T > 1.9]$. Therefore,

$P[|T| > 1.9]$ must be between $2 \times .025$ and $2 \times .05$,

that is, between .05 and .10 .

(e) With 17 d.f., the number 2.8 is between $t_{.00625} = 2.793$ and $t_{.005} = 2.898$, so $P[T > 2.8]$ must be between .005 and .00625. Because $P[|T| < 2.8] = 1 - 2P[T > 2.8]$, this probability must be between $1 - 2(.005) = .98$ and $1 - 2(.00625) = .9875$.

9.9 (a) A 98% confidence interval for μ is given by $\overline{X} \pm t_{.01}\dfrac{S}{\sqrt{n}}$ where d.f. $= n - 1$. For d.f. $= 19$ we find $t_{.01} = 2.539$. The confidence interval becomes

$$140 \pm 2.539(\frac{8}{\sqrt{20}}) = 140 \pm 4.5 \text{ or } (135.5, 144.5).$$

(b) Center $= \overline{x} = 140$. Length $= 2(4.5) = 9.0$.

(c) Usually different since the length of the interval depends on the sample standard deviation S which varies from sample to sample.

9.11 Assume a normal population. A 95% confidence interval for μ is given by $\overline{X} \pm t_{.025}\dfrac{S}{\sqrt{n}}$ where d.f. $= n - 1$. Here d.f. $= 19$ and $t_{0.25} = 2.093$. Calculating from the given data we find $\overline{x} = 137.60$ and $s = 20.143$. A 95% confidence interval for μ is then

$$137.60 \pm 2.093[\frac{20.143}{\sqrt{20}}] = 137.60 \pm 9.43 \text{ or } (\ 128.17,\ 147.03).$$

9.13 Assume a normal population. A 95% confidence interval for μ is given by $\overline{X} \pm t_{.025}\dfrac{S}{\sqrt{n}}$ where d.f. $= n - 1$. Here d.f. $= 17$ and $t_{.025} = 2.110$.

Using $\overline{x} = 3.6$ kg. and $s = .8$ kg. we get the 95% confidence interval for the mean yield:

$$3.6 \pm 2.110[\frac{.8}{\sqrt{18}}] = 3.6 \pm .4 \text{ or } (3.2, 4.0) \text{ kg.}$$

9.15 A 95% confidence interval for μ is given by $\overline{X} \pm t_{.025} \dfrac{S}{\sqrt{n}}$.

Here $n = 12$ and $t_{.025} = 2.201$ for d.f. $= 11$.

(a) A calculation of $\overline{x} \pm 2.201(\dfrac{s}{\sqrt{12}})$ has produced the result (18.6, 26.2). This interval has

$$\text{center} \quad \overline{x} = \frac{18.6 + 26.2}{2} = 22.4$$

and half-width $\quad 2.201(\dfrac{s}{\sqrt{12}}) = 22.4 - 18.6 = 3.8.$

From the last relation we find $s = 3.8[\dfrac{\sqrt{12}}{2.201}] = 5.98.$

(b) For a 98% confidence interval we use $t_{.01} = 2.718$, d.f. $= 11$. This interval is calculated as

$$\overline{x} \pm 2.718(\frac{s}{\sqrt{12}}) = 22.4 \pm 2.718(\frac{5.98}{\sqrt{12}})$$

$$= 22.4 \pm 4.7 \text{ or } (17.7, 27.1)$$

9.17 $\overline{X} \pm t_{.005}\dfrac{S}{\sqrt{n}}$ where $t_{.005} = 2.807$ for d.f. $= n - 1 = 22$. Since $\overline{x} = 5.483$ and $s = .1904$, the 99% confidence interval is calculated as

$$5.483 \pm 2.807(\dfrac{.1904}{\sqrt{23}}) = 5.483 \pm .111 \text{ or } (5.372, 5.594).$$

9.19 We are to test $H_0 : \mu = 620$ versus $H_1 : \mu \geq 620$ with $\alpha = .05$. The test statistic is $T = \dfrac{\overline{X} - 620}{S/\sqrt{10}}$. Since H_1 is right-sided, and for d.f. $= 9$ we have $t_{.05} = 1.833$, the rejection region is $R : T \geq 1.833$. Calculations from the data yield data $\overline{x} = 679.0$ and $s = 98.342$, so

$$t = \frac{679.0 - 620}{(98.342)/\sqrt{10}} = 1.90$$

which falls in the rejection region. Hence H_0 is rejected at $\alpha = .05$.

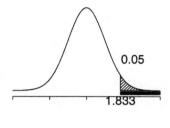

| (a) Rejection Region | (b) P-value for Problem 9.19 |

9.21 Assume that the plant heights are normally distributed.

(a) Because the claim is that μ is larger than 176 inches, we formulate

$$H_0 : \mu = 176 \text{ vs. } H_1 : \mu > 176$$

The test statistic is

$$T = \frac{\overline{X} - \mu_0}{S/\sqrt{n}} = \frac{\overline{X} - 176}{S/\sqrt{12}}, \text{ d.f. } = n - 1 = 11$$

Because H_1 is right-sided, the rejection region has the form $R : T \geq t_\alpha$.
For d.f. $= 11$, we find $t_{.05} = 1.796$ so the rejection region with $\alpha = .05$ is
$R : T \geq 1.796$.

Using the given summary statistics $\bar{x} = 182$ inches and $s = 10$ inches we
calculate

$$t = \frac{182 - 176}{10/\sqrt{12}} = 2.08$$

which is in R. Therefore, H_0 is rejected at $\alpha = .05$. We conclude that
there is strong evidence in support of the claim.

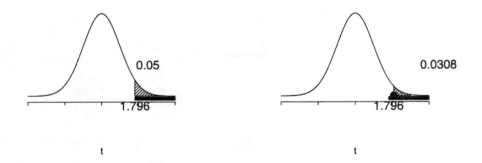

(a) Rejection Region   (b) P-value for Problem 9.21a

(b) A 95% confidence interval for μ is given by $\bar{X} \pm t_{.025} \frac{S}{\sqrt{n}}$. Here $n = 12$ so
d.f. $= 11$. We find $t_{.025} = 2.201$. Using $\bar{x} = 182$ and $s = 10$, the 95%
confidence interval for μ is calculated as

$$182 \pm 2.201 \left[\frac{10}{\sqrt{12}}\right] = 182 \pm 6 \text{ or } (176, 188) \text{ inches.}$$

9.23 A 90% confidence interval for μ is given by $\overline{X} \pm t_{.05}\frac{S}{\sqrt{n}}$. Here $n = 13$ so d.f. $= 12$, and we find $t_{.05} = 1.782$. The confidence interval is calculated as

$$26.62 \pm 1.782[\frac{6.56}{\sqrt{13}}] = 26.62 \pm 3.24$$

$$\text{or } (23.38, 29.86) \text{ feet.}$$

9.25 Since the claim is that μ is greater than 128 mm, we formulate

$$H_0 : \ \mu = 128 \text{ vs. } H_1 : \ \mu > 128$$

The test statistic is

$$T = \frac{\overline{X} - 128}{S/\sqrt{20}}, \text{ d.f. } = n - 1 = 19$$

Because H_1 is right-sided, the rejection region has the form $R : \ T \geq t_\alpha$. For d.f. $= 19, \alpha = .05$, we find $t_{.05} = 1.729$ so the rejection region is $R : \ T \geq 1.729$.

Using the summary statistics $\overline{x} = 137.60$ and $s = 20.143$ we calculate

$$t = \frac{137.60 - 128}{20.143/\sqrt{20}} = 2.13$$

which is not in R. The null hypothesis is rejected at $\alpha = .05$. Since the $P-$ value is less than .025, actual $P[T \geq 2.13] = .023$, support for the claim quite strong. The data do not show any marked departures from normality.

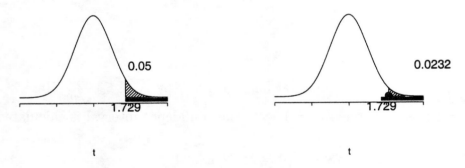

(a) Rejection Region   (b) P-value for Problem 9.25

9.27   (a) We are to test $H_0 : \ \mu = 83$ versus $H_1 : \ \mu \neq 83$ pounds. The test statistic
       is

$$T = \frac{\overline{X} - 83}{S/\sqrt{n}}, \text{ d.f. } = n - 1 = 7$$

and, with $\alpha = .05$, $t_{.025} = 2.365$. The rejection region is $R: T \leq -2.365$ or $T \geq 2.365$. Calculating from the sample data we find $\overline{x} = 73.87$, $s = 10.063$ and

$$t = \frac{73.87 - 83}{(10.063)/\sqrt{8}} = -2.566.$$

Since the observed $t = -2.57$ is in the rejection region, H_0 is rejected at $\alpha = .05$. We conclude that the mean weight is different from 83 pounds. A computer calculation gives the $P-$value .0372.

(a) Rejection Region   (b) P-value for Problem 9.27a

(b) A 95% confidence interval is given by $\overline{X} \pm t_{.025}\frac{S}{\sqrt{n}}$ where $t_{.025} = 2.365$ for d.f. $= 7$. From the given data, a 95% confidence interval for μ becomes

$$73.87 \pm 2.365[\frac{10.063}{\sqrt{8}}] = 73.87 \pm 8.41 \text{ or } (75.5, 82.3).$$

9.29 (a) Because we want to establish a decrease of the mean drying time, we
         formulate $H_0 : \mu = 90, H_1 : \mu < 90$. Assume that the drying time is
         normally distributed. The test statistic is

         $$T = \frac{\overline{X} - 90}{S/\sqrt{n}}, \text{ d.f. } = n - 1 = 14.$$

         With $\alpha = .05$, the rejection region is $R : T \leq -t_{.05} = -1.761$.

         From the given data $\overline{x} = 86, s = 4.5$ we find

         $$t = \frac{86 - 90}{(4.5)/\sqrt{15}} = -3.44.$$

         H_0 is rejected at $\alpha = .05$. It would also be rejected at $\alpha = .005$ because
         $t_{.005} = 2.977$ for d.f. $= 14$. The data strongly support the conjecture that
         $\mu < 90$.

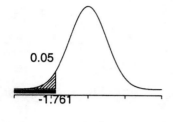
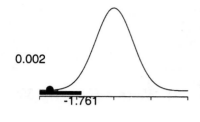

0.05

-1.761

t

0.002

-1.761

t

(a) Rejection Region   (b) P-value for Problem 9.29a

(b) A 98% confidence interval for μ is given by $\overline{X} \pm t_{.01}\dfrac{S}{\sqrt{n}}$ where $t_{.01} = 2.624$ for d.f. $= 14$. Calculating from the data, the 98% confidence interval is

$$86 \pm 2.624[\frac{4.5}{\sqrt{15}}] = 86 \pm 3 \text{ or } (83, 89).$$

9.31   (a) We test $H_0 : \mu = 300$ vs. $H_1 : \mu \neq 300$. The test statistic is

$$T = \frac{\overline{X} - 300}{S/\sqrt{n}}, \text{ d.f.} = n - 1 = 14,$$

and the rejection region is two-sided because H_1 is so.

For d.f. $= 14$, we find $t_{.025} = 2.145$ so $R : |T| \geq 2.145$.

The observed value is

$$t = \frac{340 - 300}{30/\sqrt{15}} = 5.16.$$

Since $t = 5.81$ is in the rejection region, H_0 is rejected at $\alpha = .05$. The $P-$value is very small.

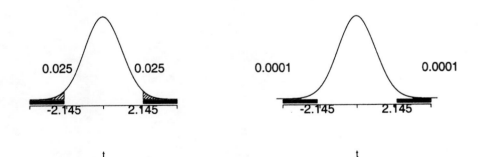

(a) Rejection Region   (b) P-value for Problem 9.31a

(b) A 95% confidence interval for μ is calculated as

$$\overline{x} \pm t_{.025}\frac{s}{\sqrt{n}} = 340 \pm 2.145[\frac{30}{\sqrt{15}}] = 340 \pm 16.6 \text{ or } (323.4, 356.6).$$

9.35 (a) Since $\mu_0 = 41$ is outside of the 90% confidence interval (38.12, 40.48), H_0 is rejected at $\alpha = .10$.

(b) Since $\mu_0 = 39$ is inside the 90% confidence interval, H_0 is not rejected at $\alpha = .10$.

9.37 Assume normal population.

(a) A 95% confidence interval for μ is given by $\overline{X} \pm t_{.025}\dfrac{S}{\sqrt{n}}$. Here $n = 8$ so d.f. $= 7$ and we find $t_{.025} = 2.365$. From the given data we calculate $\overline{x} = 6.78$ and $s = 6.58$, and obtain the confidence interval

$$6.78 \pm 2.365\left[\frac{6.58}{\sqrt{8}}\right] = 6.78 \pm 5.50 \text{ or } (1.28,\ 12.28)$$

(b) Since $\mu_0 = 15$ lies outside the 95% confidence interval, $H_0 : \ \mu = 15$ is rejected at $\alpha = .05$ in favor of $H_1 : \ \mu \neq 15$.

(c) To test $H_0 : \ \mu = 15$ vs. $H_1 : \ \mu \neq 15$, the test statistic is

$$T = \frac{\overline{X} - 15}{S/\sqrt{n}},\ \text{d.f. } = n - 1 = 7$$

Since H_1 is two-sided, the rejection region has the form $R : \ |T| \geq t_{\alpha/2}$. For d.f. $= 7$ and $\alpha = .05$ we find $\alpha/2 = .025$ and $t_{.025} = 2.365$ so the rejection region is $R : \ |T| \geq 2.365$.

Using $\overline{x} = 6.78$ and $s = 6.58$ we calculate

$$t = \frac{6.78 - 15}{(6.58)/\sqrt{8}} = -3.53$$

The observed $|t| = 3.53$ lies in R so H_0 is rejected at $\alpha = .05$. This confirms the conclusion in part (b).

9.39 The acceptance region of the $\alpha = .05$ test is

$$-1.96 \leq \frac{\overline{X} - \mu_0}{S/\sqrt{n}} \leq 1.96.$$

Rearranging the inequalities, this can also be written as

$$\overline{X} - 1.96\frac{S}{\sqrt{n}} \leq \mu_0 \leq \overline{X} + 1.96\frac{S}{\sqrt{n}}.$$

Thus, any μ_0 that lies in the interval $\overline{X} \pm 1.96\frac{S}{\sqrt{n}}$ will not be rejected at $\alpha = .05$. This interval is precisely the 95% confidence interval for μ.

9.41 (a) Since area to its left is $1 - .10 = .90$, it is the 90th percentile $\chi^2_{.10} = 23.54$

     (b) 5th percentile $\chi^2_{.95} = 12.34$

     (c) $\chi^2_{.10} = 40.26$

     (d) $\chi^2_{.95} = 3.33$.

9.43 (a) The sample standard deviation $S = 3.615$ estimates σ.

   (b) For d.f. $= n - 1 = 4$, we have $\chi^2_{.025} = 11.14$ and $\chi^2_{.975} = .48$.

      A 95% confidence interval for σ is calculated as

$$\left(3.3615\sqrt{\frac{4}{11.14}}, \;\; 3.3615\sqrt{\frac{4}{.48}}\right) = (2.01, 9.70).$$

   (c) The center of the confidence interval is

$$\frac{(2.01 + 9.70)}{2} = 5.855$$

      which is not the same as $s = 3.3615$.

9.45 We are to test $H_0 : \sigma = .6$ vs. $H_1 : \sigma < .6$. The test statistic is
$$\chi^2 = \frac{(n - 1)S^2}{(.6)^2}, \;\; \text{d.f.} = n - 1 = 39. \text{ For } \alpha = .05, \text{ the rejection region is}$$
$R : \chi^2 \leq \chi^2_{.95} = 25.7$. Using the sample value $s = .475$ we find

$$\chi^2 = \frac{39(.475)^2}{(.6)^2} = 24.4.$$

Since the observed $\chi^2 = 24.4$ lies in the rejection region, H_0 is rejected at $\alpha = .05$ in favor of $H_1 : \quad \sigma < .6$. There is strong evidence that the red pine population standard deviation is smaller than .6.

9.47 A 95% confidence interval for σ is calculated as

$$(s\sqrt{\frac{n-1}{\chi^2_{.025}}}, \quad s\sqrt{\frac{n-1}{\chi^2_{.975}}}) = (2.2706\sqrt{\frac{9}{19.02}}, \quad 2.2706\sqrt{\frac{9}{2.70}})$$

$$= (1.56, 4.15).$$

9.49 (a) For the lizard length data , $n = 20$ and the sample standard deviation $s = 20.143$. With d.f. $= 19$ we find $\chi^2_{.05} = 30.14$ and $\chi^2_{.95} = 10.12$.

A 90% confidence interval for σ is calculated as

$$\left(s\sqrt{\frac{n-1}{\chi^2_{.05}}}, \quad s\sqrt{\frac{n-1}{\chi^2_{.95}}}\right)$$

$$= \left(20.143\sqrt{\frac{19}{30.14}}, \quad 20.143\sqrt{\frac{19}{10.12}}\right) = (15.99, 27.60)$$

(b) The value $\sigma_0 = 9$ is not in the 90% confidence interval so $H_0 : \sigma = 9$ is rejected at $\alpha = .10$.

9.51 For this data set we have $n = 23$ and $s = .1904$. We find that, for d.f. $= 22$, $\chi^2_{.01} = 40.29$ and $\chi^2_{.99} = 9.54$. A 98% confidence interval for σ is then

$$\left(s\sqrt{\frac{n-1}{\chi^2_{.99}}}, \quad s\sqrt{\frac{n-1}{\chi^2_{.01}}}\right)$$

$$= \left(.1904\sqrt{\frac{22}{40.29}}, \quad .1904\sqrt{\frac{22}{9.54}}\right) = (.141, .289).$$

9.53 Assume that the population distribution is normal. Because the object is to establish the conjecture that σ is large than .4, we formulate

$$H_0 : \ \sigma = .4 \text{ vs. } H_1 : \ \sigma > .4$$

The test statistic is

$$\chi^2 = \frac{(n-1)S^2}{\sigma_0^2} = \frac{(12)S^2}{(.4)^2}, d.f. = n - 1 = 12$$

Since H_1 is right-sided, the rejection region should be of the form $R : \ \chi^2 \geq \chi_\alpha^2$. We use $\alpha = .05$, and referring to the χ^2–table for d.f. $= 12$ we find $\chi_{.05}^2 = 21.03$. The rejection region is then $R : \ \chi^2 \geq 21.03$.

Using the data, we calculate $s = .558$ and the test statistic

$$\chi^2 = \frac{12(.588)^2}{(.4)^2} = 23.35$$

Because the observed $\chi^2 = 23.35$ lies in the rejection region, H_0 is rejected at $\alpha = .05$. We conclude that there is strong evidence in support of the conjecture that $\sigma > .4$.

9.55 (a) $t_{.05}$ $= 2.353$ (c) $-t_{.05} = -2.353$

      (b) $t_{.025} = 2.179$ (d) $-t_{.05} = -1.782$

9.57 (a) From the t table we find $t_{.05} = 1.782$ for d.f. $= 12$. Now using the normal table, we interpolate to get

$$P[Z > t_{.05}] = P[Z > 1.782] = .037.$$

This probability is smaller than $.05 = P[T > t_{.05}]$.

(b) With d.f. $= 5$ we have $t_{.05} = 2.015$. Using the normal table, we find $P[Z > 2.015] = .022$ which is smaller than $.05$. With d.f. $= 20$ we have $t_{.05} = 1.725$ and the normal table gives $P[Z > 1.725] = .042$. We observe that the probability $P[Z > t_{.05}]$ is always less than $.05$ but the difference decreases with increasing d.f.'s.

9.59 For a 99% confidence interval we use $t_{.005} = 3.012$, d.f. $= 13$. The interval is

$$47 \pm 3.012 \times \frac{9.4}{\sqrt{14}} = 47 \pm 7.6 \text{ or } (39.4, 54.6).$$

9.61 A 99% confidence interval for μ is given by $\overline{X} \pm t_{.005}\dfrac{S}{\sqrt{n}}$. Here $n = 11$ so the t-distribution has d.f. $= 10$. We find $t_{.005} = 3.169$.

(a) A calculation of $\overline{x} \pm 3.169\dfrac{s}{\sqrt{11}}$ has produced the interval $(62.5, 86.9)$. This interval has

center $\overline{x} = \dfrac{62.5 + 86.9}{2} = 74.7$

half-width $3.169\dfrac{s}{\sqrt{11}} = 86.9 - 74.7 = 12.2$

so $\dfrac{s}{\sqrt{11}} = \dfrac{12.2}{3.169}$

A point estimate of μ is $\overline{x} = 74.7$

95% error margin $= t_{.025}\dfrac{s}{\sqrt{11}} = 2.228 \times \dfrac{12.2}{3.169} = 8.58$

(b) A 90% confidence interval for μ is given by $\overline{X} \pm t_{.05}\dfrac{S}{\sqrt{n}}$.

For d.f. $= 10$ we find $t_{.05} = 1.812$. Since $\overline{x} = 74.7$ and $\dfrac{s}{\sqrt{11}} = \dfrac{12.2}{3.169}$, the 90% confidence interval is calculated as

$$74.7 \pm 1.812\left[\tfrac{12.2}{3.169}\right] = 74.7 \pm 7.0 \text{ or } (67.7, 81.7).$$

9.63 We are to test $H_0 : \mu = 42$ vs. $H_1 : \mu < 42$. The test statistic is

$$T = \frac{\overline{X} - 42}{S/\sqrt{n}}, \text{ d.f. } = n - 1 = 20.$$

With $\alpha = .01$, the rejection region is $R : T \leq -t_{.01} = -2.528$.

The observed value of t is

$$t = \frac{38.4 - 42}{(5.1)/\sqrt{21}} = -3.23.$$

Since this lies in the rejection region, we reject H_0 at $\alpha = .01$. There is strong evidence that the mean time to blossom is less than 42 days. The $P-$ value, according to a computer calculation, is $P[T \leq -3.23] = .002$.

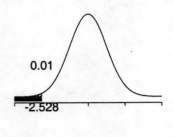

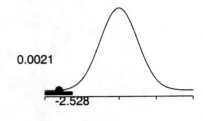

(a) Rejection Region   (b) P-value for Problem 9.63

9.65 Since H_1 is two-sided we must set a two-sided rejection region. For d.f. $= 17$, we find $t_{.01} = 2.567$ so the .02-level rejection region is $R : |T| \geq 2.567$.

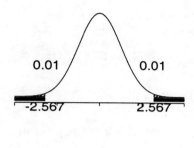

Since the observed $t = 1.59$ does not lie in this rejection region, H_0 is not rejected at $\alpha = .02$.

9.67 (a) Denote μ = mean potency after exposure. The supplier's claim is valid if $\mu > 65$, and that is to be demonstrated. We therefore formulate

$$H_0 :\ \mu = 65, \text{ vs. } H_1 :\ \mu > 65.$$

(b) Assume that the population is normal. The test statistic is

$$T = \frac{\overline{X} - 65}{S/\sqrt{n}}, \text{ d.f. } = n - 1 = 8.$$

For d.f. $= 8$, we find $t_{.05} = 1.860$ so the .05-level rejection region is

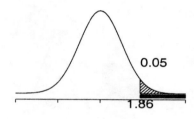

0.05

1.86

$R :\ T \geq 1.860.$ t

(c) Calculating from the sample data we find $\overline{x} = 65.22, s = 3.67$ and

$$t = \frac{65.22 - 65}{(3.67)/\sqrt{9}} = .18.$$

Since the observed $t = .18$ is not in the rejection region, H_0 is not rejected at $\alpha = .05$. In fact, it would not be rejected for any reasonable α. Therefore, the claim that $\mu > 65$ is not demonstrated.

9.69 We want to establish that $\mu > 1500$ (that is, the advertiser's claim is false). Therefore, we formulate $H_0 : \mu = 1500$, vs. $H_1 : \mu > 1500$.

Assume a normal population. The test statistic is

$$T = \frac{\overline{X} - 1500}{S/\sqrt{n}}, \text{ d.f. } = n - 1 = 4.$$

For d.f. $= 4$ we find $t_{.05} = 2.132$. To have $\alpha = .05$, we set the rejection region $R : T \geq 2.132$. Using the data $\overline{x} = 1620$ and $s = 90$, we obtain

$$t = \frac{1620 - 1500}{90/\sqrt{5}} = 2.98$$

which lies in R. Therefore, H_0 is rejected at $\alpha = .05$. Scanning the t–table for d.f. $= 4$, we find that 2.98 lies between $t_{.025} = 2.776$ and $t_{.01} = 3.747$. So the P–value is between .01 and .025 – near about .02. We conclude that the advertiser's claim is strongly contradicted.

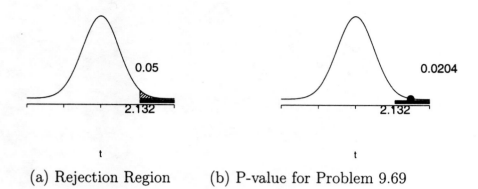

(a) Rejection Region   (b) P-value for Problem 9.69

9.71 Let μ = mean weight of males wolves.

(a) We are to test $H_0 : \mu = 83$ vs. $H_1 : \mu \neq 83$. The test statistic is

$$T = \frac{\overline{X} - 83}{S/\sqrt{n}}, d.f. = n - 1 = 10.$$

With $\alpha = .05$, the rejection region is $R : T \leq -t_{.025} = 2.228$ or $T \geq t_{.025} = 2.228$.

From the sample we calculate $\overline{x} = 91.91$ pounds and $s = 12.381$ pounds. The observed value of t is

$$t = \frac{91.91 - 83}{(12.381)/\sqrt{11}} = 2.39.$$

H_0 is rejected, at $\alpha = .05$, in favor of H_1 that the mean weight differs from 83 pounds. The P-value $P[T \leq -2.39] + P[T \geq 2.39] = .038$.

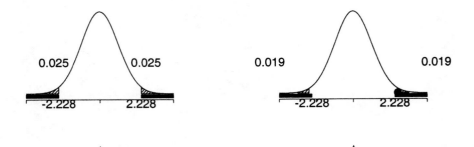

(a) Rejection Region   (b) P-value for Problem 9.71a

(b) Since $t_{.025} = 2.228$ for d.f. = 10, a 95% confidence interval for μ is calculated as

$$\overline{x} \pm t_{.025}\frac{s}{\sqrt{n}} = 91.91 \pm 2.228\left(\frac{12.381}{\sqrt{11}}\right)$$

$$= 91.91 \pm 8.32 \text{ or } (83.59, 100.23).$$

9.73   (a) $\chi^2_{.05}$ = 12.59   (c) $\chi^2_{.95}$ = 1.64

       (b) $\chi^2_{.025}$ = 38.08   (d) $\chi^2_{.975}$ = 11.69.

9.75 The population is assumed to be normal. In each case, the test statistic is

$$\chi^2 = \frac{(n-1)S^2}{1.0}, \text{ d.f.} = n - 1$$

and the rejection region has the form $R: \chi^2 \geq \chi^2_{.05}$.

(a) For d.f. $= 24$, we have $\chi^2_{.05} = 36.42$. We find

$$\chi^2 = \frac{(n-1)s^2}{1.0} = \frac{\Sigma(x_i - \overline{x})^2}{1.0} = \frac{40.16}{1.0} = 40.16$$

which is larger than 36.42. H_0 is rejected.

(b) Here $\chi^2_{.05} = 23.68$, d.f. $= 14$. The observed

$$\chi^2 = \frac{14(1.2)^2}{1.0} = 20.16 \text{ so } H_0 \text{ is not rejected.}$$

(c) We calculate $s = 2.229$ so the observed $\chi^2 = \frac{5(2.229)^2}{1.0} = 24.84$. For d.f. $= 5$ we see that $\chi^2_{.05} = 11.07$.

Hence H_0 is rejected.

9.77 A 95% confidence interval for μ is given by $\overline{S} \pm t_{.025}\frac{S}{\sqrt{n}}$. Here $n = 10$ so the t distribution has d.f. $= 9$. We find $t_{.025} = 2.262$.

(a) A calculation of $\bar{x} \pm 2.262 \frac{s}{\sqrt{10}}$ has produced the interval $(36.2, 45.8)$. This interval has

$$\text{center} \quad \bar{x} = \frac{36.2 + 45.8}{2} = 41.0,$$

$$\text{half-width} \quad 2.262 \frac{s}{\sqrt{10}} = 45.8 - 41.0 = 4.8$$

From the second relation, we get $s = 4.8 \frac{\sqrt{10}}{2.262} = 6.710$.

(b) A 98% confidence interval for μ is $\overline{X} \pm t_{.01} \frac{S}{\sqrt{n}}$.

For d.f. $= 9$ we find $t_{.01} = 2.821$. Using $s = 6.710$ calculated in part(a), the confidence interval is calculated as

$$41.0 \pm 2.821 \times \frac{6.710}{\sqrt{10}} = 41.0 \pm 6.0 \text{ or } (35.0, 47.0)$$

(c) For a 95% confidence interval for σ, we use the χ^2 table for d.f. $= 9$ to find $\chi^2_{.025} = 19.02$ and $\chi^2_{.975} = 2.70$. Using $s = 6.710$ calculated in part (a), the confidence interval is

$$\left(s\sqrt{\frac{n-1}{\chi^2_{.025}}}, \; s\sqrt{\frac{n-1}{\chi^2_{.975}}} \right) = \left(6.710\sqrt{\frac{9}{19.02}}, 6.710\sqrt{\frac{9}{2.70}} \right)$$

$$= (4.62, 12.25).$$

9.79 (a) We test $H_0 : \sigma = 20$ vs. $H_1 : \sigma < 20$. The test statistic is

$$\chi^2 = \frac{(n-1)S^2}{(20)^2}, \text{ d.f.} = n - 1 = 10,$$

and the $\alpha = .05$ rejection region is $R : \chi^2 \leq \chi^2_{.95} = 3.94$. From the data, we obtain $s = 12.381$ so The observed value of the test statistics is

$$\chi^2 = \frac{10(12.381)^2}{400} = 3.832$$

which lies in R so H_0 is rejected at $\alpha = .05$. We conclude that $\sigma < 20$.

(b) With $d.f. = 10$ we find $\chi^2_{.01} = 23.21$ and $\chi^2_{.99} = 2.56$.

A 98% confidence interval for σ is calculated as

$$(12.381\sqrt{\frac{10}{23.21}}, 12.381\sqrt{\frac{10}{2.56}}) = (8.13, 24.47).$$

Chapter 10

COMPARING TWO TREATMENTS

10.1 First group using first letter.

$$\{B,C\}, \quad \{B,E\}, \quad \{B,H\}, \quad \{B,P\}$$
$$\{C,E\}, \quad \{C,H\}, \quad \{C,P\}$$
$$\{E,H\}, \quad \{E,P\}$$
$$\{H,P\}$$

10.3 (a)

$$\{(S,G), \quad (T,E)\} \quad \{(S,G), \quad (T,R)\} \quad \{(S,G), \quad (E,R)\}$$
$$\{(S,J), \quad (T,E)\} \quad \{(S,J), \quad (T,R)\} \quad \{(S,J), \quad (E,R)\}$$
$$\{(J,G), \quad (T,E)\} \quad \{(J,G), \quad (T,R)\} \quad \{(J,G), \quad (E,R)\}$$

(b) There are 3 sets each consisting of three pairs.

10.5 (a) Point estimate of $\mu_1 - \mu_2$ has value

$$\bar{x} - \bar{y} = 73 - 66 = 7$$

Estimated $S.E. = \sqrt{\frac{s_1^2}{n_1} + \frac{s_2^2}{n_2}} = \sqrt{\frac{151}{52} + \frac{142}{44}} = 2.48$

(b) A large sample 95% confidence interval for $\mu_1 - \mu_2$ becomes

$$\bar{x} - \bar{y} \pm 1.96\sqrt{\frac{s_1^2}{n_1} + \frac{s_2^2}{n_2}} = 7 \pm 1.96(2.48) = 7 \pm 4.9 \text{ or } (2.1, 11.9)$$

10.7 We are to test $H_0 : \mu_1 - \mu_2 = 0$ vs. $H_1 : \mu_1 - \mu_2 \neq 0$.

The sample sizes are large so we employ the Z test. The test statistic is

$$Z = \frac{\overline{X} - \overline{Y}}{\sqrt{\frac{S_1^2}{n_1} + \frac{S_2^2}{n_2}}}$$

Because H_1 is two-sided, the rejection region is of the form $R: |Z| \geq z_{\alpha/2}$.

With $\alpha = .05$ we have $z_{\alpha/2} = z_{.025} = 1.96$ so the rejection region is $R: |Z| \geq 1.96$.

Calculating from the data, we find

$$z = \frac{76.4 - 81.2}{\sqrt{\frac{(8.2)^2}{90} + \frac{(7.6)^2}{100}}} = \frac{-4.8}{1.151} = -4.17$$

Because the observed z is in R, the null hypothesis is rejected at $\alpha = .05$. We conclude that the difference in the mean scores is significant at $\alpha = .05$. The $P-$value is $P[Z \leq -4.17] + P[Z \geq 4.17] = .0000$ so the evidence against no mean difference is very strong.

10.9 (a) Since the assertion is that $\mu_2 > \mu_1$, we formulate

$$H_0: \mu_1 - \mu_2 = 0 \text{ vs. } H_1: \mu_1 - \mu_2 < 0$$

(b) The sample sizes $n_1 = 78$ and $n_2 = 62$ are large so we employ the test statistic

$$Z = \frac{\overline{X} - \overline{Y}}{\sqrt{\frac{S_1^2}{n_1} + \frac{S_2^2}{n_2}}}$$

Because H_1 is left-sided, the rejection region with $\alpha = .05$ is $R: Z \leq -1.645$.

(c) The observed

$$z = \frac{89 - 108}{\sqrt{\frac{(46.2)^2}{78} + \frac{(53.4)^2}{62}}} = \frac{-19}{8.565} = -2.22$$

lies in R. Therefore, H_0 is rejected at $\alpha = .05$.

$$\text{P–value} = P[Z \leq -2.22] = .0132$$

This very small P–value signifies a strong evidence in support of H_1.

10.11 The problem here is to test $H_0 : \ \mu_1 - \mu_2 = 0$ vs. $H_1 : \ \mu_1 - \mu_2 \neq 0$ with $\alpha = .05$. Because 0 is not included in the 95% confidence interval (655, 1165), the conclusion is that H_0 is rejected at $\alpha = 1 - .95 = .05$.

10.13 The summary statistics are

$$\text{Aggressive:} \quad n_1 = 47, \quad \bar{x} = 7.92, \quad s_1 = 3.45$$

$$\text{Non-aggressive:} \quad n_2 = 38, \quad \bar{y} = 5.80, \quad s_2 = 2.87$$

Denote by μ_1 and μ_2 the population mean BPC scores for the two groups "aggressive" and "non-aggressive", respectively. Since the conjecture is that μ_1 is larger than μ_2, we formulate

$$H_0 : \ \mu_1 - \mu_2 = 0 \ \text{vs.} \ H_1 : \ \mu_1 - \mu_2 > 0$$

We employ the test statistic

$$Z = \frac{\overline{X} - \overline{Y}}{\sqrt{\dfrac{S_1^2}{n_1} + \dfrac{S_2^2}{n_2}}}$$

and note that, because H_1 is right-sided, large values of Z should be in the rejection region. Calculating from the data, we obtain

$$\text{observed } z = \frac{7.92 - 5.80}{\sqrt{\dfrac{(3.45)^2}{47} + \dfrac{(2.87)^2}{38}}} = \frac{2.12}{.06856} = 3.09$$

P–value $= P[Z \geq 3.09]$ is smaller than .001

The extremely small P–value provides a strong justification of the conjecture that μ_1 is larger than μ_2.

10.15   (a) We first obtain $\overline{x} = 7$ and $s_1^2 = (2^2 + (-2)^2 + 0^2)/2 = 8/2 = 4$. Also $\overline{y} = 6$ and $s_2^2 = (1^2 + (-2)^2 + 1^2)/2 = 6/2 = 3$. Consequently

$$s_{\text{pooled}}^2 = \frac{\Sigma(x_i - \overline{x})^2 + \Sigma(y_i - \overline{y})^2}{n_1 + n_2 - 2} = \frac{8 + 6}{3 + 3 - 2} = \frac{14}{4} = 3.5$$

(b) We estimate the common sigma by $s_{\text{pooled}} = \sqrt{3.5} = 1.87$

(c) The t statistic is

$$T = \frac{\overline{X} - \overline{Y}}{S_{\text{pooled}}\sqrt{\frac{1}{n_1} + \frac{1}{n_2}}}$$

with d.f. $= n_1 + n_2 - 2 = 3 + 3 - 2 = 4$. We calculate

$$T = \frac{7 - 6}{1.87\,\sqrt{\frac{1}{3} + \frac{1}{3}}} = .65$$

10.17 (a) We test $H_0 : \mu_1 - \mu_2 = 0$ versus $H_1 : \mu_1 - \mu_2 \neq 0$ with $\alpha = .05$. The test statistic is

$$T = \frac{\overline{X} - \overline{Y}}{S_{\text{pooled}}\sqrt{\frac{1}{n_1} + \frac{1}{n_2}}}$$

with d.f. $= n_1 + n_2 - 2$. and the rejection region has the form $R : T \leq -t_{\alpha/2}$ or $T \geq t_{\alpha/2}$ because H_1 is two-sided. With d.f. $= 8 + 11 - 2 = 17$, we find $t_{.025} = 2.110$ so the rejection region is $R : t \leq -2.110$ or $R : t \geq 2.110$. Calculating from the data we have $\overline{x} - \overline{y} = 73.9 - 91.9 = -18$, $s_1^2 = 101.268$ and $s_2^2 = 153.291$ so

$$s_{\text{pooled}} = \sqrt{\frac{(n_1 - 1)s_1^2 + (n_2 - 1)s_2^2}{n_1 + n_2 - 2}}$$

$$= \sqrt{\frac{(8-1)101.268 + (11-1)153.291}{8+11-2}} = 11.483$$

$$T = \frac{-18}{11.483\sqrt{\frac{1}{8}+\frac{1}{11}}} = -3.37$$

We reject the equality of mean weights with $\alpha = .05$. The $P-$value $P[T \leq -3.37] + P[T \geq 3.37] = .0036$, obtained from a computer calculation, greatly strengthens the conclusion.

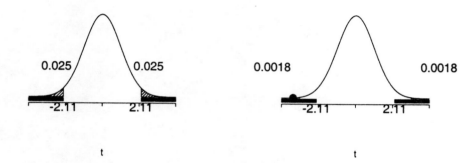

    (a) Rejection Region   (b) P-value for Problem 10.17

(b) A 95% confidence interval for $\mu_1 - \mu_2$ is given by

$$\bar{x}-\bar{y}\pm t_{.025}s_{\text{pooled}}\sqrt{\frac{1}{n_1}+\frac{1}{n_2}} = -18 \pm 2.110(11.483)\sqrt{.2159} = -18 \pm 11.3$$

or (- 29.3, - 6.7) pounds.

(c) We assume normal populations with equal σ's. The sample variances are not too different and there are no obvious outliers in the data.

10.19 The summary statistics are

$$\text{Abused:} \quad n_1 = 52, \quad \overline{x} = 2.52, \quad s_1 = 1.84$$

$$\text{Non-abused:} \quad n_2 = 67, \quad \overline{y} = 1.63, \quad s_2 = 1.22$$

Let μ_1 and μ_2 denote the population mean number of crimes for the 'abused' and 'non-abused' group, respectively.

In order to establish that μ_1 is larger than μ_2, we formulate

$$H_0 : \mu_1 - \mu_2 = 0 \text{ vs. } H_1 : \mu_1 - \mu_2 > 0$$

Because n_1 and n_2 are large, we employ the test statistic

$$Z = \frac{\overline{X} - \overline{Y}}{\sqrt{\frac{S_1^2}{n_1} + \frac{S_2^2}{n_2}}}$$

Note that large values of Z should lead to rejection of H_0 because H_1 is right-sided.

Using the summary statistics, we obtain

$$\overline{x} - \overline{y} = 2.52 - 1.63 = .89$$

$$\sqrt{\frac{s_1^2}{n_1} + \frac{s_2^2}{n_2}} = \sqrt{\frac{(1.84)^2}{52} + \frac{(1.22)^2}{67}} = .2955$$

$$Z = \frac{.89}{.2955} = 3.01$$

$$\text{P-value} = P[Z \geq 3.01] = .0013$$

This very low P-value signifies a strong evidence in support of H_1.

10.21 The summary statistics are

$$\text{Peruvian natives:} \qquad n_1 = 20, \quad \bar{x} = 46.3, \quad s_1 = 5.0$$

$$\text{U.S. subjects:} \qquad n_2 = 10, \quad \bar{y} = 38.5, \quad s_2 = 5.8$$

Denote by μ_1 and μ_2 the population mean aerobic capacity for highland natives and acclimatized lowlanders, respectively. We assume normal populations with equal σ's.

A $100(1 - \alpha)\%$ confidence interval for $\mu_1 - \mu_2$ is given by

$$\overline{X} - \overline{Y} \pm t_{\alpha/2} S_{\text{pooled}} \sqrt{\frac{1}{n_1} + \frac{1}{n_2}}$$

Here d.f $= 20 + 10 - 2 = 28, \alpha = .02, t_{.01} = 2.467,$

$$\bar{x} - \bar{y} = 46.3 - 38.5 = 7.8$$

$$s_{\text{pooled}} = \sqrt{\frac{19(5.0)^2 + 9(5.8)^2}{28}} = 5.270$$

$$s_{\text{pooled}} \sqrt{\frac{1}{n_1} + \frac{1}{n_2}} = 5.270 \sqrt{\frac{1}{20} + \frac{1}{10}} = 2.041$$

The 98% confidence interval is

$$7.8 \pm 2.467 \times 2.041 = 7.8 \pm 5.0 \text{ or } (2.8, 12.8).$$

10.23   (a) The summary statistics are calculated:

Method 1: $n_1 = 10$, $\overline{x} = 19.1$, $s_1 = 4.818$

Method 2: $n_2 = 10$, $\overline{y} = 23.3$, $s_1 = 5.559$

Denote the population mean job times corresponding to Method 1 and Method 2 by μ_1 and μ_2, respectively.

Because we are seeking strong evidence in support of $\mu_1 < \mu_2$, we formulate

$$H_0 : \ \mu_1 - \mu_2 = 0 \ \text{vs.} \ H_1 : \ \mu_1 - \mu_2 < 0$$

Assuming normal populations with equal σ's, the test statistic is

$$T = \frac{\overline{X} - \overline{Y}}{S_{\text{pooled}}\sqrt{\frac{1}{n_1} + \frac{1}{n_2}}}, \ \text{d.f.} = n_1 + n_2 - 2$$

Because H_1 is left-sided, the rejection region has the form $R : \ T \leq -t_\alpha$. Here $\alpha = .05$, d.f. $= 10+10-2 = 18$, $t_{.05} = 1.734$ so we set $R : \ T \leq -1.734$.

Using the summary statistics we calculate

$$\overline{x} - \overline{y} = 19.1 - 23.3 = -4.2$$

$$s_{\text{pooled}} = \sqrt{\frac{9(4.818)^2 + 9(5.559)^2}{18}} = 5.201$$

$$s_{\text{pooled}}\sqrt{\frac{1}{n_1} + \frac{1}{n_2}} = 5.201\sqrt{\frac{1}{10} + \frac{1}{10}} = 2.326$$

$$t = \frac{-4.2}{2.326} = -1.81$$

Since $t = -1.81$ lies in R, H_0 is rejected at $\alpha = .05$. We conclude that the mean job time is significantly less for Method 1 than Method 2.

(b) See (a).

(c) A $100(1 - \alpha)\%$ confidence interval for $\mu_1 - \mu_2$ is given by

$$\overline{X} - \overline{Y} \pm t_{\alpha/2} S_{\text{pooled}}\sqrt{\frac{1}{n_1} + \frac{1}{n_2}}$$

Here $\alpha = .05$, d.f. $= 18$, $t_{.025} = 2.101$. Using the calculations in part (a), we obtain the 95% confidence interval

$$-4.2 \pm 2.101 \times 2.326 = -4.2 \pm 4.89 \text{ or } (-9.09, .69).$$

10.25 The summary statistics are

$$\text{Control:} \quad n_1 = 6, \quad \overline{x} = 41.8, \quad s_1 = 7.6$$

$$\text{Hormone treated:} \quad n_2 = 6, \quad \overline{y} = 60.8, \quad s_2 = 16.4$$

Let μ_1 and μ_2 denote the population mean weight gains for the control group and the treated group, respectively. For demonstrating that μ_2 is higher than μ_1 we formulate

$$H_0 : \ \mu_1 - \mu_2 = 0 \text{ vs. } H_1 : \ \mu_1 - \mu_2 < 0$$

The sample sizes are small. We assume that the populations are both normal. However, the assumption of equal σ's is not reasonable as s_1 and s_2 are far apart (s_2/s_1 is larger than 2). Therefore, we use the conservative test without pooling. The test statistic is

$$T^* = \frac{\overline{X} - \overline{Y}}{\sqrt{\frac{S_1^2}{n_1} + \frac{S_2^2}{n_2}}}, \qquad \text{d.f.} = \text{smaller of } n_1 - 1 \text{ and } n_2 - 1$$

Here, d.f. $= 5$, and we find $t_{.05} = 2.015$. Since H_1 is left-sided, we set the rejection region with $\alpha = .05$ as $R : T^* \leq -2.015$. Using the summary statistics we calculate

$$\overline{x} - \overline{y} = 41.8 - 60.8 = -19.0$$

$$\sqrt{\frac{s_1^2}{n_1} + \frac{s_2^2}{n_2}} = \sqrt{\frac{(7.6)^2}{6} + \frac{(16.4)^2}{6}} = 7.379$$

$$t^* = \frac{-19.0}{7.379} = -2.57$$

Since the observed $t^* = -2.57$ lies in R, we reject H_0 at $\alpha = .05$. We conclude that the hormone treated rats have a significantly higher weight gain.

10.27 The summary statistics are

Isometric method: $\quad n_1 = 10, \quad \overline{x} = 2.4, \quad s_1 = 0.8$

Isotonic method: $\quad n_2 = 10, \quad \overline{y} = 3.2, \quad s_2 = 1.0$

(a) Let μ_1 and μ_2 denote the population mean decrease in abdomen measurements under the isometric method and the isotonic method, respectively, To demonstrate that the isotonic method is more effective, that is $\mu_2 > \mu_1$ we formulate

$$H_0 : \ \mu_1 - \mu_2 = 0 \text{ vs. } H_1 : \ \mu_1 - \mu_2 < 0$$

We assume that both populations are normal and $\sigma_1 = \sigma_2$.

The test statistic is

$$T = \frac{\overline{X} - \overline{Y}}{S_{\text{pooled}}\sqrt{\frac{1}{n_1} + \frac{1}{n_2}}}, \ \text{d.f.} = n_1 + n_2 - 2$$

Because H_1 is left-sided, the rejection region has the form $R : \ T \leq -t_\alpha$.
Taking $\alpha = .05$ and noting that, with d.f. $= 10 + 10 - 2 = 18, t_{.05} = 1.734$,
we set $R : \ T \leq -1.734$.

Using the summary statistics we calculate

$$\overline{x} - \overline{y} = 2.4 - 3.2 = -.8$$

$$s_{\text{pooled}} = \sqrt{\frac{9(.8)^2 + 9(1.0)^2}{18}} = .906$$

$$s_{\text{pooled}}\sqrt{\frac{1}{n_1} + \frac{1}{n_2}} = .906\sqrt{\frac{1}{10} + \frac{1}{10}} = .405$$

$$t = \frac{-.8}{.405} = -1.98$$

Since the observed t lies in R, H_0 is rejected at $\alpha = .05$. The superiority
of the isotonic method is demonstrated by the data.

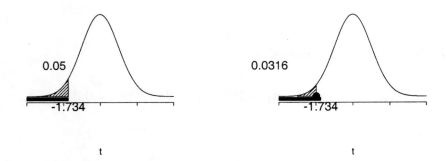

(a) Rejection Region   (b) P-value for Problem 10.27

(b) A 95% confidence interval for $\mu_1 - \mu_2$ is calculated as

$$-.8 \pm 2.101 \times .405 = -.8 \pm .85 \text{ or } (-1.65, .05) \text{ centimeters.}$$

10.29 (a) The summary statistics are

$$n_1 = 14, \quad \overline{x} = 28.1, \quad s_1 = 3.6$$

$$n_2 = 12, \quad \overline{y} = 30.0, \quad s_2 = 5.1$$

Assume normal populations with equal σ's. A $100(1 - \alpha)\%$ confidence interval for $\mu_1 - \mu_2$ is given by

$$\overline{X} - \overline{Y} \pm t_{\alpha/2} S_{\text{pooled}} \sqrt{\frac{1}{n_1} + \frac{1}{n_2}}$$

For a 98% confidence interval, we have $\alpha/2 = .01$. With d.f. $= 14 + 12 - 2 = 24$, we find $t_{.01} = 2.492$. The summary statistics yield

$$\overline{x} - \overline{y} = -1.9$$

$$S_{\text{pooled}} = \sqrt{\frac{13(3.6)^2 + 11(5.1)^2}{24}} = 4.352$$

$$S_{\text{pooled}} \sqrt{\frac{1}{n_1} + \frac{1}{n_2}} = 4.352 \sqrt{\frac{1}{14} + \frac{1}{12}} = 1.712$$

so the confidence interval is

$$-1.9 \pm 2.492 \times 1.712 = -1.9 \pm 4.27 \text{ or } (-6.17, 2.37)$$

(b) The summary statistics are

$$n_1 = 120, \quad \overline{x} = 410, \quad s_1 = 26$$

$$n_2 = 91, \quad \overline{y} = 390, \quad s_2 = 38$$

Since the sample sizes are large, the assumption of normal populations is not needed. Neither is the assumption of equal variances. A $100(1 - \alpha)\%$ confidence interval is given by

$$\overline{X} - \overline{Y} \pm z_{\alpha/2} \sqrt{\frac{S_1^2}{n_1} + \frac{S_2^2}{n_2}}$$

We calculate

$$\overline{x} - \overline{y} = 410 - 390 = 20$$

$$\sqrt{\frac{s_1^2}{n_1} + \frac{s_2^2}{n_2}} = \sqrt{\frac{(26)^2}{120} + \frac{(38)^2}{91}} = 4.637$$

$$z_{.01} = 2.33$$

so the 98% confidence interval is

$$20 \pm 2.33 \times 4.637 = 20 \pm 10.8 \text{ or } (9.2, 30.8)$$

(c) The summary statistics are

$$n_1 = 11, \quad \overline{x} = 1.25, \quad s_1 = 0.079$$

$$n_2 = 14, \quad \overline{y} = 1.32, \quad s_2 = .326$$

The sample sizes are small. We assume that both populations are normal. The assumption of equal σ's is not reasonable because s_2/s_1 is large than 2. A conservative $100(1 - \alpha)\%$ confidence interval for $\mu_1 - \mu_2$ is given by

$$\overline{X} - \overline{Y} \pm t_{\alpha/2}^* \sqrt{\frac{S_1^2}{n_1} + \frac{S_2^2}{n_2}}, \quad \text{d.f.} = \text{smaller of } n_1 - 1 \text{ and } n_2 - 1$$

We calculate

$$\overline{x} - \overline{y} = -.07$$

$$\sqrt{\frac{s_1^2}{n_1} + \frac{s_2^2}{n_2}} = .0903$$

For d.f. $= 10, t_{.01} = 2.764$

so the 98% confidence interval is

$$-.07 \pm 2.764 \times .0903 = -.07 \pm .250 \text{ or } (-.320, .180)$$

(d) The summary statistics are

$$n_1 = 65, \quad \overline{x} = 75.6, \quad s_1 = 18.1$$

$$n_2 = 55, \quad \overline{y} = 62.5, \quad s_2 = 6.8$$

The sample sizes are large. We do not need the assumption of normal populations or equal variances. A 98% confidence interval for $\mu_1 - \mu_2$ is found (by the same method as in part (b)) to be

$$13.1 \pm 2.33 \times 2.425 = 13.1 \pm 5.65 \text{ or } (7.45, 18.75).$$

10.31

```
Two sample T for WolfFwt vs WolMwt

              N      Mean     StDev    SE Mean
WolfFwt       8      73.9     10.1       3.6
WolMwt       11      91.9     12.4       3.7


97% CI for mu WolfFwt - mu WolMwt: ( -30.7,  -5.4)
T-Test mu WolfFwt = mu WolMwt (vs not =): T= -3.38 P=0.0036 DF=17
Both use Pooled StDev = 11.5
```

10.33 We drew slips with α, β, τ so group 1 is {alpha, beta, tau}.

10.35 We must be careful. Probably mothers who are warmer toward everyone have a much higher tendency to nurse their babies than mothers who have colder personalities.

10.37 (a) The $n = 6$ paired differences $d = x - y$ are

$$2, 2, 4, 0, 1, -3$$

Their mean and standard deviation are

$$\bar{d} = 1, \quad s_D = 2.366$$

$$\text{so } t = \frac{1}{(2.366)/\sqrt{6}} = 1.035$$

(b) d.f. $= n - 1 = 5$

10.39 It is a matched pair sample because there may be considerable variation of conditions in the different plants. The paired differences $d =$ (before $-$ after) are $2, 1, -1, 2, 3, -1$. We assume these differences constitute a random sample from a normal distribution with mean δ. The null hypothesis of no change is $H_0 : \delta = 0$, and the alternative of more loss before than after is $H_1 : \delta > 0$. We calculate $\bar{d} = 1.0, s_D = 1.673, s_D/\sqrt{6} = .683, t = 1.0/.683 = 1.46$, d.f. $= 5$. With $\alpha = .05$, the rejection region is $T > t_{.05} = 2.015$. The observed t being less than 2.015, H_0 is not rejected. The claim of effectiveness of the safety program is not demonstrated.

10.41 This is a matched pair design because from each farm a pair of milk specimens are taken, and then one is treated with PC while the other is not.

(a) The paired differences

$$d = \text{(with PC} - \text{without PC) are}$$

$$7, 6, -2, 10, 0, 8, 4$$

which produce the summary statistics

$$n = 7, \quad \bar{d} = 4.714, \quad s_D = 4.348$$

We assume that the population distribution of the D's is normal, and denote $\delta =$ the population mean. Because the conjecture is that the mean response with PC is higher than without PC, that is $\delta > 0$, we formulate

$$H_0 : \ \delta = 0 \text{ vs. } H_1 : \ \delta > 0.$$

The test statistic is

$$T = \frac{\overline{D}}{S_D/\sqrt{n}}, \ \text{d.f.} = n - 1$$

and the rejection region is of the form $R : \ T \geq t_\alpha$ because H_1 is right-sided. With d.f. $= 6$ we find $t_{.05} = 1.943$ so the rejection region with $\alpha = .05$ is $R : T \geq 1.943$.

Using the summary statistics, we calculate

$$t = \frac{4.714}{(4.348)/\sqrt{7}} = 2.87$$

The observed $t = 2.87$ is in R so H_0 is rejected at $\alpha = .05$.

We conclude that there is strong evidence in support of the conjecture.

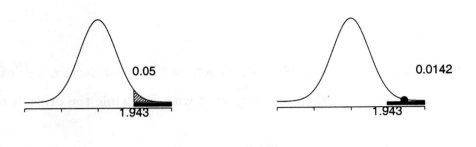

(a) Rejection Region   (b) P-value for Problem 10.41

(b) A 90% confidence interval for δ is

$$\bar{d} \pm t_{.05} s_D / \sqrt{n}$$

$$= 4.71 \pm 1.943 \times \frac{4.35}{\sqrt{7}}$$

$$= 4.71 \pm 3.19 \text{ or } (1.52, \ 7.90).$$

10.43 This is a matched pair design. The paired differences $A - B$ are

$$5, -1, 2, 3, 0, -1, 4, 3, -3.$$

We assume these differences to be a random sample from a normal distribution with mean δ. The null hypothesis of no difference between the methods is then $H_0 : \delta = 0$. The alternative is $H_1 : \delta \neq 0$. To apply the t test, we calculate

$$\overline{d} = 1.333, \quad s_D = 2.693$$

$$t = \frac{\overline{d}}{s_D/\sqrt{n}} = \frac{1.333}{2.693/\sqrt{9}} = 1.48, \text{ d.f.} = 8$$

The alternative being two sided, the rejection region is $R : |T| \geq t_{.025} = 2.306$ for $\alpha = .05$. The observed t is smaller, so H_0 is not rejected at $\alpha = .05$. The difference is not significant.

10.45 This is a matched-pair design because the two plots in the same farm are alike in respect of soil condition, rainfall, temperature and other factors that affect the yield. We calculate the paired differences $d = $ yield of $A-$ yield of B

$$5, 6, -2, 9, 0, 9, 7, 2$$

and the summary statistics

$$n = 8, \quad \bar{d} = 4.50, \quad s_D = 4.11$$

Assume that the d's form a random sample from a normal population whose mean we denote by δ.

(a) The object is to substantiate that A has a higher mean yield than B, that is $\delta > 0$. We therefore formulate

$$H_0 : \delta = 0 \text{ vs. } H_1 : \delta > 0$$

The test statistic is

$$T = \frac{\overline{D}}{S_D/\sqrt{n}}, \text{ d.f.} = n - 1$$

and the rejection region has the form $R : T \geq t_\alpha$ because H_1 is right-sided. With d.f. $= 7$ we find $t_{.05} = 1.895$ so the rejection region with $\alpha = .05$ is $R : T \geq 1.895$. Calculating from the data we have

$$t = \frac{4.50}{4.11/\sqrt{8}} = 3.10$$

which is in R. Hence, H_0 is rejected at $\alpha = .05$. We conclude that A has a significantly higher mean yield than B.

(b) At each farm, the assignment of the two plots to strain A and strain B should be randomized. Label the two plots 1 and 2, toss a coin and if a head turns up assign plot 1 to strain A, 2 to B. If a tail turns up assign plot 1 to B and 2 to A. Repeat this process for all the farms.

10.47 We calculate the paired differences $d = (\text{left} - \text{right})$

$$2, 3, 15, -2, -1, 1, -1, 7, 2, 10$$

and the summary statistics

$$n = 10, \bar{d} = 3.60, s_D = 5.461$$

Assume that the population distribution of the paired differences is normal with a mean denoted by δ.

(a) We are to test $H_0 : \delta = 0$ vs. $H_1 : \delta > 0$

The test statistic is

$$T = \frac{\bar{D}}{S_D/\sqrt{n}}, \text{ d.f.} = n - 1$$

Using $\alpha = .05$ we set the right-sided rejection region $R : T \geq 1.833$ because with d.f. $= 9$ we have $t_{.05} = 1.833$.

Using the summary statistics we find

$$t = \frac{3.60}{(5.461)/\sqrt{10}} = 2.085$$

which lies in R. Therefore, H_0 is rejected at $\alpha = .05$.

(b) With $1 - \alpha = .9$, we have $\alpha/2 = .05$ and $t_{.05} = 1.833$, d.f. $= 9$. The required confidence interval for δ is $\bar{d} \pm t_{.05} s_D/\sqrt{n} = 3.6 \pm 1.833 \times 1.727 = 3.6 \pm 3.17$ or $(.43, 6.77)$.

10.49 (a) $\hat{p}_1 - \hat{p}_2 = .5 - .65 = -.15$

Estimated $S.E. = \sqrt{\dfrac{\hat{p}_1 \hat{q}_1}{n_1} + \dfrac{\hat{p}_2 \hat{q}_2}{n_2}}$

$\qquad = \sqrt{\dfrac{.5 \times .5}{100} + \dfrac{.65 \times .35}{200}} = .060$

A 95% confidence interval for $p_1 - p_2$ is

$$-.15 \pm 1.96 \times .060 = -.15 \pm .12 \text{ or } (-.27, -.03).$$

(b) The test statistic is

$$Z = \frac{\hat{p}_1 - \hat{p}_2}{\sqrt{\hat{p}\hat{q}}\sqrt{\dfrac{1}{n_1} + \dfrac{1}{n_2}}}$$

Because H_1 is left-sided, small values of Z should lead to rejection of H_0.

We calculate

$$\text{Pooled estimate } \hat{p} = \frac{n_1 \hat{p}_1 + n_2 \hat{p}_2}{n_1 + n_2} = \frac{50 + 130}{300} = .60$$

$$z = \frac{-.15}{\sqrt{.60 \times .40}\sqrt{\dfrac{1}{100} + \dfrac{1}{200}}} = \frac{-.15}{.0600} = -2.50$$

The significance probability of the observed z is

P–value $= P[Z \leq -2.50] = .0062$

This means that H_0 would be rejected with α as small as .0062.

10.51 (a) The problem here is to test

$$H_0 : p_A = p_B \text{ versus } H_1 : p_A < p_B$$

The test statistic is

$$Z = \frac{\hat{p}_1 - \hat{p}_2}{\sqrt{\hat{p}\hat{q}}\sqrt{\frac{1}{n_1} + \frac{1}{n_2}}}$$

where $\hat{p}_1$ is identified as $\hat{p}_A$ and $\hat{p}_2$ as $\hat{p}_B$.

Since H_1 is left-sided and $z_{.05} = 1.645$, the rejection region is set as $R : Z \leq -1.645$.

We calculate

$$\hat{p}_A = \frac{51}{120} = .425, \quad \hat{p}_B = \frac{88}{150} = .587$$

$$\text{Pooled estimate } \hat{p} = \frac{51 + 88}{120 + 150} = \frac{139}{270} = .515$$

$$z = \frac{.425 - .587}{\sqrt{.515 \times .485}\sqrt{\frac{1}{120} + \frac{1}{150}}} = \frac{-.162}{.0612} = -2.65$$

The observed $z = -2.65$ lies in R so H_0 is rejected at $\alpha = .05$. The P-value $P[Z \leq -2.65] = .004$. There is very strong evidence that drug B has a higher cure rate than drug A.

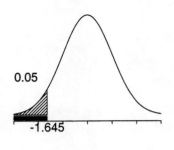

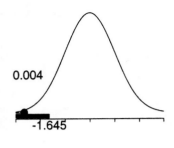

(a) Rejection Region   (b) P-value for Problem 10.51

(b) $\hat{p}_B - \hat{p}_A = .587 - .425 = .162$

Estimate $S.E. = \sqrt{\dfrac{.425 \times .575}{120} + \dfrac{.587 \times .413}{150}} = .060$

A 95% confidence interval for $p_B - p_A$ is

$$.162 \pm 1.96 \times .060 = .162 \pm .120 \text{ or } (.04, .28).$$

10.53 Letting p_1 and p_2 denote the probabilities of 'resistant' for the HRL and LRL
groups, respectively, the problem is to test

$$H_0: \; p_1 = p_2 \text{ versus } H_1: \; p_1 < p_2.$$

The test statistic is

$$Z = \frac{\hat{p}_1 - \hat{p}_2}{\sqrt{\hat{p}\hat{q}}\sqrt{\frac{1}{n_1} + \frac{1}{n_2}}}$$

Small values of Z should lead to the rejection of H_0 since H_1 is left-sided.

We calculate

$$\hat{p}_1 = \frac{15}{49} = .306, \quad \hat{p}_2 = \frac{42}{54} = .778$$

$$\text{Pooled estimate } \hat{p} = \frac{15 + 42}{49 + 54} = \frac{57}{103} = .553$$

$$z = \frac{.306 - .778}{\sqrt{.553 \times .447}\sqrt{\frac{1}{49} + \frac{1}{54}}} = -4.81$$

The significance probability of this observed z is

$$\text{P–value} = P[Z \leq -4.81] < .0001$$

The extremely small P–value signifies a strong evidence in support of H_1.

10.55 Let p_1 and p_2 denote the probability of being a chronic offender for abused and non-abused groups, respectively. We are to test

$$H_0: \ p_1 = p_2 \text{ vs. } H_1: \ p_1 > p_2$$

The test statistic is

$$Z = \frac{\hat{p}_1 - \hat{p}_2}{\sqrt{\hat{p}\hat{q}}\sqrt{\frac{1}{n_1} + \frac{1}{n_2}}}$$

Since H_1 is right-sided and $z_{.01} = 2.33$, the rejection region with $\alpha = .01$ is set as $R : Z \geq 2.33$.

We calculate

$$\hat{p}_1 = \frac{21}{85} = .247, \quad \hat{p}_2 = \frac{11}{120} = .092$$

$$\text{Pooled estimate } \hat{p} = \frac{21 + 11}{85 + 120} = \frac{32}{205} = .156$$

$$z = \frac{.247 - .092}{\sqrt{.156 \times .844}\sqrt{\frac{1}{85} + \frac{1}{120}}} = 3.01$$

The observed $z = 3.01$ lies in R so H_0 is rejected at $\alpha = .01$.

We conclude that there is strong evidence in support of the conjecture.

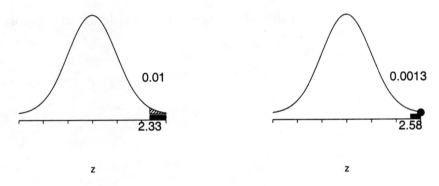

(a) Rejection Region   (b) P-value for Problem 10.55

10.57 Denote the probability of survival by

p_1 for the treated group (with carbolic acid)

p_2 for the control group (without carbolic acid)

We are to test $H_0 : p_1 = p_2$ vs. $H_1 : p_1 \neq p_2$.

The test statistic is

$$Z = \frac{\hat{p}_1 - \hat{p}_2}{\sqrt{\hat{p}\hat{q}}\sqrt{\frac{1}{n_1} + \frac{1}{n_2}}}$$

Since H_1 is two-sided, and $z_{.025} = 1.96$, the rejection region with $\alpha = .05$ is set as $R : |Z| \geq 1.96$.

We calculate

$$\hat{p}_1 = \frac{34}{40} = .850, \quad \hat{p}_2 = \frac{19}{35} = .543$$

$$\text{Pooled estimate } \hat{p} = \frac{34 + 19}{40 + 35} = \frac{53}{75} = .707$$

$$z = \frac{.850 - .543}{\sqrt{.707 \times .293}\sqrt{\frac{1}{40} + \frac{1}{35}}} = 2.91$$

The observed $z = 2.91$ lies in R so H_0 is rejected at $\alpha = .05$.

$$\text{P-value} = P[|Z| \geq 2.91] = 2P[Z \leq -2.91]$$

$$= 2 \times .0018 = .0036$$

This means that H_0 would be rejected with α as small as .0036. A difference in the survival rates is strongly demonstrated by the data.

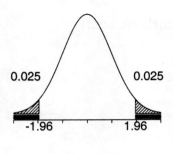 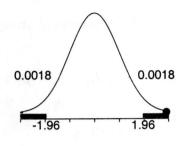

(a) Rejection Region   (b) P-value for Problem 10.57

10.59 Denote the population proportion of ≤ 8 hours of sleep by

p_1 for the age group 30–40, and

p_2 for the age group 60–70.

We are to test $H_0 : p_1 = p_2$ vs. $H_1 : p_1 > p_2$.

The test statistic is

$$Z = \frac{\hat{p}_1 - \hat{p}_2}{\sqrt{\hat{p}\hat{q}}\sqrt{\frac{1}{n_1} + \frac{1}{n_2}}}$$

Since H_1 is right-sided, large values of Z should lead to rejection of H_0.

We calculate

$$\hat{p}_1 = \frac{173}{250} = .692, \quad \hat{p}_2 = \frac{120}{250} = .480$$

Pooled estimate $\hat{p} = \frac{293}{500} = .586$

$$z = \frac{.692 - .480}{\sqrt{.586 \times .414}\sqrt{\frac{1}{250} + \frac{1}{250}}} = \frac{.212}{.044} = 4.81$$

P–value $= P[|Z| \geq 4.81] = 2P[Z \leq -4.81]$

(The exact result cannot be determined from the table but we can say that it is less than $2 \times .0002 = .0004$)

In view of an extremely small P–value, we conclude that there is strong support of H_1.

10.61 (a) Denote the probability of having prominent wrinkles by

p_1 for smokers, and

p_2 for non-smokers.

We are to test $H_0 : p_1 = p_2$ vs. $H_1 : p_1 > p_2$.

The test statistic is

$$Z = \frac{\hat{p}_1 - \hat{p}_2}{\sqrt{\hat{p}\hat{q}}\sqrt{\frac{1}{n_1} + \frac{1}{n_2}}}$$

Because H_1 is right-sided, large values of Z should lead to rejection of H_0.

We calculate

$$\hat{p}_1 = \frac{95}{150} = .633, \quad \hat{p}_2 = \frac{103}{250} = .412$$

Pooled estimate $\hat{p} = \dfrac{95 + 103}{150 + 250} = \dfrac{198}{400} = .495$

$$z = \frac{.633 - .412}{\sqrt{.495 \times .505}\sqrt{\frac{1}{150} + \frac{1}{250}}} = \frac{.221}{.0516} = 4.28$$

P–value $= P[Z \geq 4.28]$, cannot be exactly determined from the table but we can say that it is less than .0002.

In view of an extremely small P–value, we conclude that the conjecture is strongly substantiated by the data.

(b) A direct causal relation between smoking and wrinkled skin cannot be readily concluded. Various psycho-physiological factors could influence both the smoking habit and presence of wrinkled skin.

10.63 (a) Denote the probability of getting hepatitis by

p_1 for the 'vaccine' group, and

p_2 for the 'placebo' group

We are to test $H_0 : p_1 = p_2$ vs. $H_1 : p_1 < p_2$

The test statistic is

$$Z = \frac{\hat{p}_1 - \hat{p}_2}{\sqrt{\hat{p}\hat{q}}\sqrt{\frac{1}{n_1} + \frac{1}{n_2}}}$$

We take $\alpha = .01$. Noting that H_1 is left-sided and $z_{.01} = 2.33$, the rejection region is set as $R : Z \leq -2.33$.

We calculate

$$\hat{p} = \frac{11}{549} = .020, \quad \hat{p}_2 = \frac{70}{534} = .131$$

Pooled estimate $\hat{p} = \frac{11 + 70}{1083} = .075$

$$z = \frac{.020 - .131}{\sqrt{.075 \times .925}\sqrt{\frac{1}{549} + \frac{1}{534}}} = \frac{-.111}{.016} = -6.93$$

The observed $z = -6.93$ lies in R so H_0 is rejected at $\alpha = .01$.

The data strongly demonstrate that the vaccine is effective.

(b) $\hat{p}_2 - \hat{p}_1 = .131 - .020 = .111$

Estimated $S.E. = \sqrt{\frac{\hat{p}_1\hat{q}_1}{n_1} + \frac{\hat{p}_2\hat{q}_2}{n_2}}$

$$= \sqrt{\frac{.020 \times .980}{549} + \frac{.131 \times .869}{534}} = .016$$

A 95% confidence interval for $p_2 - p_1$ is

$$.111 \pm 1.96 \times .016 = .111 \pm .031 \text{ or } (.08, .14).$$

10.65 (a) $\hat{p}_E = \frac{78}{150} = .520, \quad \hat{p}_H = \frac{39}{160} = .244$

$\hat{p}_E - \hat{p}_H = .276$

Estimated $S.E. = \sqrt{\frac{.520 \times .480}{150} + \frac{.244 \times .756}{160}} = .053$

A 95% confidence interval for $p_E - p_H$ is

$.276 \pm 1.96 \times .053 = .276 \pm .104$ or $(.17, .38)$.

(b) $\hat{p}_H = \frac{39}{160} = .244, \quad \hat{p}_C = \frac{43}{200} = .215$

$\hat{p}_H - \hat{p}_C = .029$

Estimated $S.E. = \sqrt{\frac{.244 \times .756}{160} + \frac{.215 \times .785}{200}} = .045$

A 90% confidence interval for $p_H - p_C$ is

$.029 \pm 1.645 \times .045 = .029 \pm .074$ or $(-.045, .103)$

(c) A 95% confidence interval for a population proportion is given by

$$\hat{p} \pm 1.96\sqrt{\frac{\hat{p}\hat{q}}{n}}$$

Calculations for the individual groups are as follows:

Diabetes. $n = 160, \hat{p}_D = 41/160 = .256$

Confidence interval for p_D :

$.256 \pm 1.96\sqrt{\frac{.256 \times .744}{160}}$

$= .256 \pm .068$ or $(.19, .32)$

Heart condition. $n = 160, \hat{p}_H = 39/160 = .244$

Confidence interval for p_H:

$.244 \pm .067$ or $(.18, .31)$

Epilepsy. $n = 150, \hat{p}_E = 78/150 = .520$

Confidence interval for p_E :

$$.520 \pm .080 \text{ or } (.44, .60)$$

<u>Control</u>. $n = 200, \hat{p}_C = 43/200 = .215$

Confidence interval for p_C :

$$.215 \pm .057 \text{ or } (.16, .27).$$

10.67 For each case we are to test

$$H_0 : \mu_1 - \mu_2 = 0 \text{ vs. } H_1 : \mu_1 - \mu_2 \neq 0$$

at $\alpha = .05$. The sample sizes are $n_1 = n_2 = 52$

The test statistic is

$$Z = \frac{\overline{X} - \overline{Y}}{\sqrt{\frac{S_1^2}{n_1} + \frac{S_2^2}{n_2}}}$$

Because H_1 is two-sided, $\alpha/2 = .025$ and $z_{.025} = 1.96$, the rejection region is $R : |Z| \geq 1.96$.

(a) We have

$$|\overline{x} - \overline{y}| = 8$$
$$\sqrt{\frac{s_1^2}{n_1} + \frac{s_2^2}{n_2}} = \sqrt{\frac{(20)^2}{52} + \frac{(28)^2}{52}} = 4.772$$
$$|z| = \frac{8}{4.772} = 1.68$$

Since 1.68 is not in R, H_0 is not rejected. The difference is not significant at $\alpha = .05$.

(b) We have

$$|\overline{x} - \overline{y}| = 8$$
$$\sqrt{\frac{s_1^2}{n_1} + \frac{s_2^2}{n_2}} = \sqrt{\frac{(12)^2}{52} + \frac{(15)^2}{52}} = 2.664$$
$$|z| = \frac{8}{2.664} = 3.00$$

Since 3.00 is in R, H_0 is rejected. The difference is significant at $\alpha = .05$.

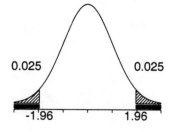

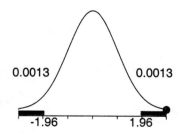

(a) Rejection Region   (b) P-value for Problem 10.67

10.69 (a) Because the claim is that $\mu_2 > \mu_1 + 10$, we formulate

$$H_0 : \mu_2 - \mu_1 = 10 \text{ vs. } H_1 : \mu_2 - \mu_1 > 10$$

(b) The test statistic is

$$Z = \frac{(\overline{Y} - \overline{X}) - 10}{\sqrt{\frac{S_1^2}{n_1} + \frac{S_2^2}{n_2}}}$$

and with $\alpha = .10$, the rejection region is $R : Z \geq z_{.10} = 1.28$

(c) We calculate

$$\overline{y} - \overline{x} = 27.84 - 16.21 = 11.63$$

$$z = \frac{11.63 - 10}{.7721} = 2.11$$

Since 2.11 is in R, H_0 is rejected at $\alpha = .10$.

$$\text{P--value} = P[Z \geq 2.11] = .0174$$

H_0 would be rejected with α as small as .0174. This signifies a strong evidence in support of H_1.

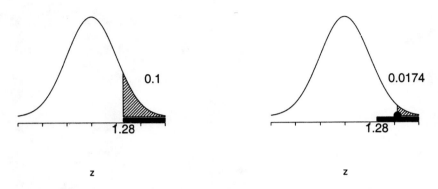

(a) Rejection Region   (b) P-value for Problem 10.69

10.71   (a) The 90% confidence interval for $\mu_A - \mu_B$ is calculated as

$$\bar{x} - \bar{y} \pm z_{.05}\sqrt{\frac{s_1^2}{n_1} + \frac{s_2^2}{n_2}} = 4.64 - 4.03 \pm 1.645\sqrt{\frac{(1.25)^2}{55} + \frac{(1.82)^2}{58}}$$

$$= .61 \pm .48 \text{ or } (.13, 1.09).$$

We are 90% confident that μ_A is .13 to 1.09 hours longer than μ_B.

(b) The 95% confidence interval for μ_A is

$$\bar{x} \pm z_{.025} \frac{s_1}{\sqrt{n_1}} = 4.64 \pm 1.96 \frac{(1.25)}{\sqrt{55}}$$

$$= 4.64 \pm .33 \text{ or } (4.31, 4.97).$$

10.73 (a) We calculate $\bar{x} = 8, \Sigma(x - \bar{x})^2 = 14, \bar{y} = 5$ and $\Sigma(y - \bar{y})^2 = 20$.

Consequently,

$$s^2_{\text{pooled}} = \frac{\Sigma(x - \bar{x})^2 + \Sigma(y - \bar{y})^2}{n_1 + n_2 - 2} = \frac{14 + 20}{5 + 4 - 2} = 4.857.$$

(b) $$t = \frac{(\bar{x} - \bar{y}) - 2}{s_{\text{pooled}} \sqrt{\frac{1}{n_1} + \frac{1}{n_2}}} = \frac{(8 - 5) - 2}{\sqrt{4.857}\sqrt{\frac{1}{5} + \frac{1}{4}}} = .68$$

d.f. $= n_1 + n_2 - 2 = 7$.

10.75 The individual 95% confidence intervals are calculated as

$$\mu_1 : \quad \overline{x} \pm t_{.025}\frac{s_1}{\sqrt{n_1}} = 74.3 \pm 2.262\frac{4.8}{\sqrt{10}} \text{ or } (70.87,\ 77.73) \text{ minutes.}$$

$$\mu_2 : \quad \overline{y} \pm t_{.025}\frac{s_2}{\sqrt{n_2}} = 81.6 \pm 2.201\frac{6.5}{\sqrt{12}} \text{ or } (77.47,\ 85.73) \text{ minutes.}$$

10.77 We are to test $H_0 : \ \mu_1 - \mu_2 = 0$ vs. $H_1 : \ \mu_1 - \mu_2 > 0$.

The test statistic is

$$T = \frac{\overline{X} - \overline{Y}}{S_{\text{pooled}}\sqrt{\frac{1}{n_1} + \frac{1}{n_2}}}, \ \text{d.f.} = n_1 + n_2 - 2$$

Noting that H_1 is right-sided, d.f. $= 22$, $t_{.05} = 1.717$ we set the rejection region $R : T \geq 1.717$ with $\alpha = .05$.

Using the calculations in Exercise 10.76 we have

$$t = \frac{1.1}{.5499} = 2.00$$

which lies in R. Hence, H_0 is rejected at $\alpha = .05$. In the t–table we further see that for d.f $= 22$, $t_{.025} = 2.074$ so the P–value here is between .025 and .05. We conclude that there is strong evidence in support of H_1.

10.79 Noting that, with d.f. $= 11$, $t_{.025} = 2.201$ and using the calculations in Exercise 10.78, we obtain the 95% confidence interval for $\mu_1 - \mu_2$:

$16 \pm 2.201 \times 12.848 = 16 \pm 28.3$ or $(-12.3, 44.3)$.

10.81 The summary statistics are

City A: $n_1 = 75$, $\bar{x} = 37.8$, $s_1 = 6.8$

City B: $n_2 = 100$, $\bar{y} = 43.2$, $s_2 = 7.5$

(a) Let μ_1 and μ_2 denote the population mean age for City A and City B, respectively. To demonstrate a difference between μ_1 and μ_2 we formulate

$$H_0 : \mu_1 - \mu_2 = 0 \text{ vs. } H_1 : \mu_1 - \mu_2 \neq 0$$

Because the sample sizes are large, we employ the test statistic

$$Z = \frac{\overline{X} - \overline{Y}}{\sqrt{\dfrac{S_1^2}{n_1} + \dfrac{S_2^2}{n_2}}}$$

Noting that H_1 is two-sided, $\alpha = .02, z_{.01} = 2.33$, the rejection region is $R : |Z| \geq 2.33$. We calculate

$$z = \frac{37.8 - 43.2}{\sqrt{\dfrac{(6.8)^2}{75} + \dfrac{(7.5)^2}{100}}} = -4.97$$

Since the observed z is in R, H_0 is rejected at $\alpha = .02$. There is strong evidence that the mean ages are different.

(b) We calculate the 98% confidence interval as

$$\bar{x} - \bar{y} \pm z_{.01}\sqrt{\frac{s_1^2}{n_1} + \frac{s_2^2}{n_2}} = 37.8 - 43.2 \pm 2.33\sqrt{\frac{(6.8)^2}{75} + \frac{(7.5)^2}{100}}$$

$$= -5.4 \pm 2.53 \text{ or } (-7.93, -2.87).$$

We are 98% confident the mean age for City B is 2.87 to 7.93 years higher than for City A.

(c) The individual 98% confidence intervals are calculated as

$$\mu_1 : \bar{x} \pm 2.33\frac{s_1}{\sqrt{n_1}} = 37.8 \pm 2.33\frac{6.8}{\sqrt{75}} \text{ or } (35.97, 39.63) \text{ years.}$$

$$\mu_2 : \bar{y} \pm 2.33\frac{s_2}{\sqrt{n_2}} = 43.2 \pm 2.33\frac{7.5}{\sqrt{100}} \text{ or } (41.45, 44.95) \text{ years.}$$

10.83 (a) The experimental units are the 16 cars. Since the gasoline could vary considerably in different cars, a matched pair design is appropriate. With each car, observe the mileage with a tank-full of each brand of gasoline. Randomize the order of the two brands independently for each car.

(b) A matched pair design would be effective. Use the two brands, one on each half of a board. Randomize between the two halves. Each half of a board is an experimental unit.

(c) Each girl is an experimental unit. Since they are homogeneous with regard to age, the design of independent samples will be appropriate for comparing the two teaching methods. Divide the 40 girls at random in two groups of 20 each (for an operational method, see Exercise 2.26 (c)). Use method 1 for one group and method 2 for the other.

10.85 The data come from a matched pair design with each location giving a pair. We calculate differences $d = $ (with additive – without additive)

$$3, 3, -1, -1, 3, -1, 4, 2, 4, 4$$

and the summary statistics

$$n = 10, \quad \bar{d} = 2.0, \quad s_D = 2.160$$

Assume that the paired differences are normally distributed and denote the population mean by δ. Because an increased growth with the additive means that $\delta > 0$, we formulate

$$H_0 : \delta = 0 \text{ vs. } H_1 : \delta > 0$$

The test statistic is

$$T = \frac{\overline{D}}{S_D/\sqrt{n}}, \text{ d.f.} = n - 1$$

Because H_1 is right-sided, d.f. $= 9$, $t_{.05} = 1.833$, the rejection region with $\alpha = .05$ is set as $R : T \geq 1.833$.

We calculate

$$t = \frac{2.0}{(2.160)/\sqrt{10}} = 2.93$$

The observed $t = 2.93$ is in R so H_0 is rejected at $\alpha = .05$. There is strong evidence that the additive is effective.

10.87   (a) For each location, toss a coin. If head appears, choose East for the additive
            and if tail appears, choose West.

        (b) W W E W E E W E E E

10.89 This is a case of independent samples. We calculate the summary statistics

Standard: $n_1 = 8$, $\quad \overline{x} = 69.125$, $\quad s_1 = 4.086$

New: $\quad n_2 = 8$, $\quad \overline{y} = 75.375$, $\quad s_2 = 4.373$

Assume normal populations with equal σ's. Let μ_1 and μ_2 denote the population mean scores for the standard method and the new method, respectively.

A $100(1 - \alpha)\%$ confidence interval for $\mu_1 - \mu_2$ is given by

$$\overline{X} - \overline{Y} \pm t_{\alpha/2} S_{\text{pooled}} \sqrt{\frac{1}{n_1} + \frac{1}{n_2}}$$

For a 95% confidence interval, we have $\alpha/2 = .025$. With d.f. $= 8 + 8 - 2 = 14$, we find $t_{.025} = 2.145$. We calculate

$$\overline{x} - \overline{y} = -6.250$$

$$S_{\text{pooled}} = \sqrt{\frac{7(4.086)^2 + 7(4.373)^2}{14}} = 4.232$$

$$S_{\text{pooled}} \sqrt{\frac{1}{n_1} + \frac{1}{n_2}} = 4.232 \sqrt{\frac{1}{8} + \frac{1}{8}} = 2.116$$

A 95% confidence interval for the difference of means is calculated as

$$-6.25 \pm 2.145 \times 2.116 = -6.25 \pm 4.54 \text{ or } (-10.79, -1.71)$$

Since the confidence interval does not include 0, the null hypothesis $H_0 : \mu_1 - \mu_2 = 0$ will be rejected in favor of $H_1 : \mu_1 - \mu_2 \neq 0$, at $\alpha = .05$. Thus, a significant difference of the means is indicated.

10.91 Let p_1 denote the probability of hail from the seeded clouds and p_2, the probability of hail from the clouds that are not seeded. We are to test

$$H_0 : \ p_1 = p_2 \text{ vs. } H_1 : \ p_1 < p_2$$

The test statistic is

$$Z = \frac{\hat{p}_1 - \hat{p}_2}{\sqrt{\hat{p}\hat{q}}\sqrt{\frac{1}{n_1} + \frac{1}{n_2}}}$$

Since H_1 is left-sided, small values of Z should lead to rejection of H_0.

We calculate

$$\hat{p}_1 = \frac{7}{50} = .140, \quad \hat{p}_2 = \frac{43}{165} = .261$$

Pooled estimate $\hat{p} = \frac{50}{215} = .233$

$$z = \frac{.140 - .261}{\sqrt{.233 \times .767}\sqrt{\frac{1}{50} + \frac{1}{165}}} = \frac{-.121}{.0682} = -1.77$$

P–value $= P[Z \leq -1.77] = .0384$

H_0 would be rejected with α as small as .0384. We conclude that there is fairly strong evidence in support of the conjecture.

10.93 Let

$$\mu_1 = \text{mean of males}$$

$$\mu_2 = \text{mean for females.}$$

A computer calculation gives the summary statistics

$$n_1 = 5 \quad \overline{x} = 13.40 \quad s_1 = 8.82$$

$$n_2 = 11 \quad \overline{y} = 13.09 \quad s_2 = 5.52$$

The 95 percent confidence interval for $\mu_1 - \mu_2$ uses $\alpha/2 = .025$. With d.f. $= 5 + 11 - 2 = 14$, we find $t_{.025} = 2.145$. We calculate

$$\overline{x} - \overline{y} = .31$$

$$s_{\text{pooled}} = \sqrt{\frac{4(8.82)^2 + 10(5.52)^2}{14}} = 6.63$$

$$s_{\text{pooled}}\sqrt{\frac{1}{n_1} + \frac{1}{n_2}} = 6.63\sqrt{\frac{1}{5} + \frac{1}{11}} = 3.58$$

The 95% confidence interval is

$$.31 \pm 2.145 \times 3.58 = .31 \pm 7.68 \text{ or } (-7.4, 8.0)$$

The confidence interval includes 0 so this is one plausible value for $\mu_1 - \mu_2$. Males should have much more testosterone than females so we expected the difference of means to have a large positive value. The males are not healthy.

10.95 (a) Denote by p_1 and p_2 the population proportions of uremic and normal patients, respectively, who are allergic to the antibiotic. We are to test $H_0: p_1 = p_2$ vs. $H_1: p_1 > p_2$.

The test statistic is

$$Z = \frac{\hat{p}_1 - \hat{p}_2}{\sqrt{\hat{p}\hat{q}}\sqrt{\frac{1}{n_1} + \frac{1}{n_2}}}$$

With $\alpha = .01$, the rejection region is $R : Z \geq 2.33$.

We calculate

$$\hat{p}_1 = 38/100 = .38, \quad \hat{p}_2 = 21/100 = .21$$

Pooled estimate $\hat{p} = (38 + 21)/(100 + 100) = .295$.

$$z = \frac{.38 - .21}{\sqrt{.295 \times .705}\sqrt{\frac{1}{100} + \frac{1}{100}}} = \frac{.17}{.0645} = 2.64.$$

The observed $z = 2.64$ lies in R so H_0 is rejected at $\alpha = .01$. The data provide strong evidence of a higher incidence of allergy in uremic patients.

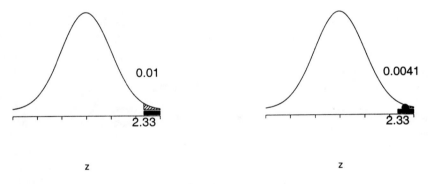

(a) Rejection Region   (b) P-value for Problem 10.95

(b) For a 95% confidence interval of $p_1 - p_2$, we calculate

$$\hat{p}_1 - \hat{p}_2 = .17$$

$$\text{Estimated } S.E. = \sqrt{\frac{.38 \times .62}{100} + \frac{.21 \times .79}{100}} = .063.$$

The required confidence interval is

$$.17 \pm 1.96 \times .063 = .17 \pm .12 \text{ or } (.05, .29).$$

10.97 Let p_1 and p_2 denote the probabilities of a high degree of hopelessness among boys and girls, respectively.

We are to test

$$H_0 : p_1 = p_2 \text{ vs. } H_1 : p_1 < p_2.$$

The test statistic is

$$Z = \frac{\hat{p}_1 - \hat{p}_2}{\sqrt{\hat{p}\hat{q}}\sqrt{\frac{1}{n_1} + \frac{1}{n_2}}}$$

Because H_1 is left-sided, small values of Z should lead to rejection of H_0.

We calculate

$$\hat{p}_1 = \frac{6}{105} = .057, \quad \hat{p}_2 = \frac{12}{105} = .114$$

Pooled estimate $\hat{p} = \dfrac{6 + 12}{105 + 105} = \dfrac{18}{210} = .086$

$$z = \frac{.057 - .114}{\sqrt{.086 \times .914}\sqrt{\frac{1}{105} + \frac{1}{105}}} = \frac{-.057}{.0387} = -1.47$$

P–value $= P[Z \leq -1.47] = .0708$

An unreasonably high α of .0708 is required for rejecting H_0. We conclude that a higher prevalence of hopelessness among girls is not demonstrated by the data.

Chapter 11

Regression Analysis - I
Simple Linear Regression

11.1 The points on the line $y = 2 + 3x$ for $x = 1$ and $x = 4$ are (1, 5) and (4, 14) correspondingly. The intercept is 2 and the slope is 3.

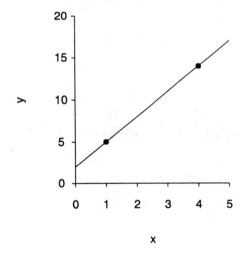

11.3 The model is $Y = \beta_0 + \beta_1 x + e$ where $E(e) = 0$ and $sd(e) = \sigma$, so $\beta_0 = 7$, $\beta_1 = -5$ and $\sigma = 3$.

11.5 (a) $Y = \beta_0 + \beta_1 x + e = 3 - 4x + e$ with $E(e) = 0$ and $sd(e) = \sigma$.

   At $x = 1$, $E(Y) = 3 - 4(1) = -1$ and $sd(Y) = sd(e) = 4$.

   (b) At $x = 2$, $E(Y) = 3 - 4(2) = -5$ and $sd(Y) = 4$.

11.7 The straight line for the means of the model $Y = 7 + 2x + e$ is $y = 7 + 2x$. The line is shown below.

11.9 (a) At $x = 4$, $E(Y) = \beta_0 + \beta_1(4) = 4 + 3(4) = 16$.

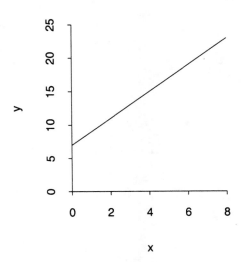

Exercise 11.7. The line $y = 7 + 2x$.

At $x = 5$, $E(Y) = \beta_0 + \beta_1(5) = 4 + 3(5) = 19$.

(b) No, only the mean is higher. By chance the error e at $x = 5$, which has standard deviation 4, could be quite negative and/or the error at $x = 4$ very large.

11.13 (a) The residuals and their sum are calculated in the following table.

x	y	$\hat{y} = 4.845 - .615x$	$\hat{e} = y - \hat{y}$	$(y - \hat{y})^2$
0	5	4.845	0.155	0.0240
1	4	4.230	−0.230	0.0529
6	1	1.155	−0.155	0.0240
3	3	3.000	0.0	0.0
5	2	1.770	0.230	0.0529
			0.0	0.1538
			$\sum(y - \hat{y})$	SSE

(b) $SSE = $ Sum of squares residuals $= .1538$

$$SSE = S_{yy} - \frac{S_{xy}^2}{S_{xx}} = 10 - \frac{(-16)^2}{26} = .1538 \quad \text{(check)}$$

(c) $S^2 = \dfrac{SSE}{n-2} = \dfrac{.1538}{5-2} = .0513$

11.15   (a) The calculations of $\bar{x}, \bar{y}, S_{xx}, S_{xy}$ and S_{yy} are given in the following table.

x	y	$x - \bar{x}$	$y - \bar{y}$	$(x - \bar{x})(y - \bar{y})$	$(x - \bar{x})^2$	$(y - \bar{y})^2$
0	2	-2	-2	4	4	4
1	1	-1	-3	3	1	9
2	4	0	0	0	0	0
3	5	1	1	1	1	1
4	8	2	4	8	4	16
10	20	0	0	16	10	30
				S_{xy}	S_{xx}	S_{yy}

$\bar{x} = 2$ $\bar{y} = 4$

(b) $\hat{\beta}_1 = S_{xy}/S_{xx} = 16/10 = 1.6,$ and $\hat{\beta}_0 = 4 - (1.6)(2) = .8$

(c) The fitted line is $\hat{y} = .8 + 1.6x$.

11.17   (a) $\hat{\beta}_1 = \dfrac{S_{xy}}{S_{xx}} = \dfrac{2.677}{10.82} = .2474,$ and $\hat{\beta}_0 = 2.32 - .2474(3.5) = 1.454$

The fitted line is $\hat{y} = 1.454 + .247x$.

(b) $SSE = S_{yy} - \dfrac{S_{xy}^2}{S_{xx}} = 1.035 - \dfrac{(2.677)^2}{10.82} = .373$

(c) $s^2 = \dfrac{SSE}{n-2} = \dfrac{.373}{12} = .031$

11.19 We first calculate the means and sums of squares and products.

$$n = 7 \quad \overline{x} = 889/7 = 127 \quad \overline{y} = 520/7 = 74.286$$

$$\textstyle\sum_{i=1}^{7} x_i^2 = 113,237 \quad \sum_{i=1}^{7} y_i^2 = 39,328$$

$$\textstyle\sum_{i=1}^{7} x_i y_i = 66,392$$

$$S_{xx} = \sum_{i=1}^{7} x_i^2 - (\sum_{i=1}^{7} x_i)^2/7 = 113,237 - (889)^2/7 = 334$$

$$S_{yy} = \sum_{i=1}^{7} y_i^2 - (\sum_{i=1}^{7} y_i)^2/7 = 39,328 - (520)^2/7 = 699.43$$

$$S_{xy} = \sum_{i=1}^{7} x_i y_i - (\sum_{i=1}^{7} x_i)(\sum_{i=1}^{7} y_i)/7 = 66,392 - (520)(889)/7 = 352$$

(a) $\hat{\beta}_1 = \dfrac{S_{xy}}{S_{xx}} = \dfrac{352}{334} = 1.054,$ and $\hat{\beta}_0 = 74.286 - \dfrac{352}{334}(127) = -59.56$

The fitted line is $\hat{y} = -59.56 + 1.054x$.

(b) $SSE = S_{yy} - \dfrac{S_{xy}^2}{S_{xx}} = 699.43 - \dfrac{(352)^2}{334} = 328.46$

(c) $s^2 = \dfrac{SSE}{n-2} = \dfrac{328.46}{5} = 65.69$

11.21 Since $\hat{\beta}_1 = \dfrac{S_{xy}}{S_{xx}}$ and $SSE = S_{yy} - \dfrac{S_{xy}^2}{S_{xx}}$

(a) $SSE = S_{yy} - \dfrac{S_{xy}^2}{S_{xx}} = S_{yy} - \dfrac{S_{xy}}{S_{xx}} S_{xy} = S_{yy} - \hat{\beta}_1 S_{xy}$

(b) $SSE = S_{yy} - \hat{\beta}_1 S_{xy} = S_{yy} - \hat{\beta}_1 S_{xy} \dfrac{S_{xx}}{S_{xx}} = S_{yy} - \hat{\beta}_1^2 S_{xx}$

11.23 (a) At x_i, $\hat{y}_i = \hat{\beta}_0 + \hat{\beta}_1 x_i = (\bar{y} - \hat{\beta}_1 \bar{x}) + \hat{\beta}_1 x_i = \bar{y} + \hat{\beta}_1 (x_i - \bar{x})$.

(b) The residual at x_i is $y_i - \hat{y}_i$ or

$$\hat{e}_i = y_i - \bar{y} - \hat{\beta}_1 (x_i - \bar{x}) = (y_i - \bar{y}) - \hat{\beta}_1 (x_i - \bar{x}).$$

Summing the $\hat{e}_i$ we obtain $\sum (y_i - \bar{y}) - \hat{\beta}_1 \sum (x_i - \bar{x}) = 0 + 0$ because the sum of deviations $\sum (y_i - \bar{y})$ and $\sum (x_i - \bar{x})$ are zero.

(c) $e_i^2 = (y_i - \bar{y})^2 - 2\hat{\beta}_1 (x_i - \bar{x})(y_i - \bar{y}) + \hat{\beta}_1^2 (x_i - \bar{x})^2$

Summing, we obtain

$$SSE = \sum \hat{e}_i^2 = S_{yy} - 2\hat{\beta}_1 S_{xy} + \hat{\beta}_1^2 S_{xx} = S_{yy} - 2\dfrac{S_{xy}}{S_{xx}} S_{xy} + \dfrac{S_{xy}^2}{S_{xx}^2} S_{xx} = S_{yy} - \dfrac{S_{xy}^2}{S_{xx}}$$

11.25 A 90% confidence interval for β_0 is calculated as

$$\hat{\beta}_0 \pm 2.353s\sqrt{\frac{1}{n} + \frac{\bar{x}^2}{S_{xx}}} = 4.845 \pm 2.353\sqrt{.0513}\sqrt{\frac{1}{5} + \frac{3^2}{26}} = (4.45,\ 5.24)$$

11.27   (a) The calculations are carried out in the table.

x	y	$x - \bar{x}$	$y - \bar{y}$	$(x - \bar{x})(y - \bar{y})$	$(x - \bar{x})^2$	$(y - \bar{y})^2$
0	.9	-2	-1.6	3.2	4	2.56
1	2.1	-1	$-.4$	.41	1	.16
2	2.4	0	$-.1$	0	0	.01
3	3.3	1	.8	.83	1	.64
4	3.8	2	1.3	2.6	4	1.69
10	12.5	0	0	7.0	10	5.06
$\bar{x} = 2$ $\bar{y} = 2.5$				S_{xy}	S_{xx}	S_{yy}

$$\hat{\beta}_1 = .7, \quad \text{and} \quad \hat{\beta}_0 = 2.0 - .7(3) = -.1$$

$$SSE = S_{yy} - \frac{S_{xy}^2}{S_{xx}} = 5.06 - \frac{49}{10} = .16$$

$$s^2 = \frac{.16}{5 - 2} = .0533.$$

(b) We test $H_0 : \beta_1 = 1$ versus $H_1 : \beta_1 \neq 1$ at the 5% level. The rejection region is $R : |T| > t_{.025} = 3.182$ for 3 degrees of freedom. Since

$$\frac{\hat{\beta}_1 - 1}{s/\sqrt{S_{xx}}} = \frac{.7 - 1}{\sqrt{.0533/10}} = -4.109$$

we reject $H_0 : \beta_1 = 1$ at the 5% level.

(c) The expected value is estimated by $-.1 + .7(3.5) = 2.35$ and

$$s\sqrt{\frac{1}{n} + \frac{(3.5 - \bar{x})^2}{S_{xx}}} = \sqrt{.0533}\sqrt{\frac{1}{5} + \frac{(3.5 - 2)^2}{10}} = .225$$

$2.85 \pm 3.182(.110)$ or $(2.50, 3.20)$

(d) $\hat{\beta}_0 \pm 2.353s\sqrt{\frac{1}{n} + \frac{\bar{x}^2}{S_{xx}}} = .4 \pm 2.353\sqrt{.053}\sqrt{\frac{1}{5} + \frac{9}{10}} = (-.17, .97)$

11.29 (a) The assessed value is $x = 90$ in thousands of dollars. At $x = 90$,

$$\hat{y} = 15.5 + .869x = 15.5 + .869(90) = 93.71.$$

Also,

$$s\sqrt{\frac{1}{n} + \frac{(90 - \bar{x})^2}{S_{xx}}} = 2.314\sqrt{\frac{1}{7} + \frac{(90 - 95.657)^2}{1644.317}} = .932$$

A 95% confidence interval for the mean selling price of all homes assessed at \$90,000 is $93.71 \pm 2.571(.932)$ or $(91.31, \; 96.11)$.

(b) $s\sqrt{1 + \frac{1}{n} + \frac{(90 - \bar{x})^2}{S_{xx}}} = 2.314\sqrt{1 + \frac{1}{7} + \frac{(90 - 95.657)^2}{1644.317}} = 2.495$

so a 95% confidence interval for the selling price of an individual home assessed at \$90,000 is $93.71 \pm 2.571(2.495)$ or $(87.30, \; 100.12)$.

11.31   (a) $2.846 + 1.595(1.2) \pm (2.160)\sqrt{0.116}\sqrt{\frac{1}{15} + \frac{(1.2 - 1.1)^2}{4.2}} = (4.57, \ 4.95)$

(b) $2.846 + 1.595(1.5) \pm (2.160)\sqrt{0.116}\sqrt{\frac{1}{15} + \frac{(1.5 - 1.1)^2}{4.2}} = (5.00, \ 5.48)$

11.33   (a) At $x = 4$, the predicted mean response is

$$994 + 0.10373(5000) = 1512.7$$

(b) Since $t_{.05} = 1.701$ with 28 d.f., a 90% confidence interval for the mean response at $x = 5000$ is calculated as

$$1512.7 \pm (1.701)s\sqrt{\frac{1}{n} + \frac{(5000 - \bar{x})^2}{S_{xx}}}$$

$$= 1512.7 \pm (1.701)\sqrt{89636}\sqrt{\frac{1}{30} + \frac{(5000 - 8,354)^2}{97,599,296}}$$

$$= (1316.4, \ 1709.0)$$

11.35 (a) At $x = 3$, the predicted mean response is

$$.3381 + .83099(3) = 2.8311.$$

(b) Since $t_{.05} = 1.714$ with 23 d.f., a 90% confidence interval for the mean response at $x = 3$ is calculated as

$$2.8311 \pm (1.714)s\sqrt{\frac{1}{n} + \frac{(3 - \bar{x})^2}{S_{xx}}}$$

$$= 2.8311 \pm (1.714)\sqrt{.0146}\sqrt{\frac{1}{25} + \frac{(3 - 1.793)^2}{1.848}} = (2.643,\ 3.020)$$

(c) A 90% confidence interval is calculated as

$$.338 + .831(2) \pm (1.714)\sqrt{.0146}\sqrt{\frac{1}{25} + \frac{(2 - 1.793)^2}{1.848}} = (1.95,\ 2.05)$$

11.37 (a) We first calculate the means and sums of squares and products.

$$n = 11 \quad \bar{x} = 313.5/11 = 28.5 \quad \bar{y} = 29/11 = 2.636$$

$$\sum_{i=1}^{11} x_i^2 = 8986.37 \qquad \sum_{i=1}^{11} y_i^2 = 95$$

$$\sum_{i=1}^{11} x_i y_i = 852.7$$

$$S_{xx} = \sum_{i=1}^{11} x_i^2 - (\sum_{i=1}^{11} x_i)^2/11 = 8986.37 - (313.5)^2/11 = 51.62$$

$$S_{yy} = \sum_{i=1}^{11} y_i^2 - (\sum_{i=1}^{11} y_i)^2/11 = 95 - (29)^2/11 = 18.545$$

$$S_{xy} = \sum_{i=1}^{11} x_i y_i - (\sum_{i=1}^{11} x_i)(\sum_{i=1}^{11} y_i)/11 = 852.7 - (313.5)(29)/11 = 26.2$$

Then, $\hat{\beta}_1 = \dfrac{S_{xy}}{S_{xx}} \dfrac{26.2}{51.62} = .5076$, and $\hat{\beta}_0 = 2.636 - .5076 \times 28.5 = -11.83$

The fitted line is $\hat{y} = -11.83 + .5076x$. Further, $SSE = S_{yy} - \dfrac{S_{xy}^2}{S_{xx}} =$

$$18.545 - \frac{(26.2)^2}{51.62} = 5.247 \text{ so}$$

$$s^2 = \frac{SSE}{n-2} = \frac{5.247}{9} = .583$$

(b) With $\alpha = .05$, we test $H_0 : \beta_1 = 0$ versus $H_1 : \beta_1 \neq 0$. The rejection region is $R : T < -t_{.025}$ or $T > t_{.025} = 2.262$ for $11 - 2 = 9$ degrees of freedom. Since

$$t = \frac{\hat{\beta}_1}{s/\sqrt{S_{xx}}} = \frac{.5076}{\sqrt{.583}/\sqrt{51.62}} = 4.78$$

so we reject $H_0 : \beta_1 = 0$ in favor of $H_1 : \beta_1 \neq 0$. The P–value is smaller than .005.

(c) Since $t_{.05} = 1.833$ for 9 degrees of freedom and $x^* = 27$, the confidence interval is

$$\hat{\beta}_0 + \hat{\beta}_1 x^* \pm t_{.05} s \sqrt{\frac{1}{n} + \frac{(x^* - \overline{x})^2}{S_{xx}}}$$

$$-11.83 + .5076(27) \pm 1.833\sqrt{.583}\sqrt{\frac{1}{11} + \frac{(27 - 28.5)^2}{51.62}}$$

$$= \quad 1.875 \pm .513 \quad \text{or} \quad (1.36, 2.39)$$

(d)

$$-11.83 + .5076(27) \pm 1.833\sqrt{.583}\sqrt{1 + \frac{1}{11} + \frac{(27 - 28.5)^2}{51.62}}$$

$$= \quad 1.875 \pm 1.491 \quad \text{or} \quad (0.38, 3.37)$$

11.39 (a) $\bar{x} = 2$, $\bar{y} = 239.8$, $S_{xx} = 10$, $S_{yy} = 4952.8$, $S_{xy} = 222$

$\hat{y} = 195.4 + 22.2x$

The scatter diagram and the fitted line is shown in Figure 11.5.

(b) With $\alpha = .05$, we test $H_0 : \beta_1 = 0$ versus $H_1 : \beta_1 > 0$. The rejection region is $R : T > t_{.05} = 2.353$ for 3 degrees of freedom. Since

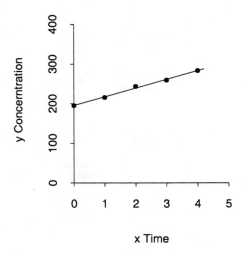

Figure 11.5: Exercise 11.39. Scatter diagram and the fitted line.

$$t = \frac{\hat{\beta}_1}{s/\sqrt{S_{xx}}} = \frac{22.2}{\sqrt{24.4}/\sqrt{10}} = 14.2$$

we reject $H_0 : \beta_1 = 0$ in favor of $H_1 : \beta_1 > 0$. The P–value is smaller than .0005.

(c) $x^* = 1999 - 1976 = 23$ so the estimate is $195.4 + 22.2(23) = 706.0$.

The straight-line relation may not hold to year 1999 or $x^* = 23$.

11.41 (a) The data are set in columns C1 to C3.

```
            NAME C2 'ATTITUDE' C3 'ANXIETY'
            REGRESS C3 ON 1 PREDICTOR IN C2

        THE REGRESSION EQUATION I2424242424242424242424S
        ANXIETY = - 0.361 + 1.11 ATTITUDE

        PREDICTOR      COEF       STDEV    T-RATIO        P
        CONSTANT     -0.3614     0.4120     -0.88     0.387
        ATTITUDE      1.1124     0.1443      7.71     0.000

        S = 0.4071      R-SQ = 64.3%     R-SQ(ADJ) = 63.2%
```

```
ANALYSIS OF VARIANCE

SOURCE        DF        SS        MS        F        P
REGRESSION     1      9.8548    9.8548    59.47    0.000
ERROR         33      5.4687    0.1657
TOTAL         34     15.3235
```

(b) Using the subscripts in C1, we unstack the data in columns C2 and C3 into two blocks: the data for female students in C11 and C12 and male students in C13 and C14.

```
        UNSTACK (C2 C3) INTO (C11 C12) (C13 C14);
        SUBSCRIPTS IN C1.
        NAME C13 'ATTITUDE' C14 'ANXIETY'
        REGRESS C14 ON 1 PREDICTOR IN C13

THE REGRESSION EQUATION IS
ANXIETY = 0.441 + 0.857 ATTITUDE

PREDICTOR      COEF      STDEV     T-RATIO       P
CONSTANT      0.4408    0.7273      0.61      0.552
ATTITUDE      0.8566    0.2449      3.50      0.003

S = 0.4160      R-SQ = 40.5%      R-SQ(ADJ) = 37.1%

ANALYSIS OF VARIANCE

SOURCE        DF        SS        MS        F        P
REGRESSION     1      2.1165    2.1165    12.23    0.003
ERROR         18      3.1149    0.1731
TOTAL         19      5.2315
```

(c) The data for female students are set in C11 and C12.

```
        NAME C11 'ATTITUDE' C12 'ANXIETY'
        REGRESS C12 ON 1 PREDICTOR IN C11

THE REGRESSION EQUATION IS
```

ANXIETY = - 0.699 + 1.22 ATTITUDE

PREDICTOR	COEF	STDEV	T-RATIO	P
CONSTANT	-0.6990	0.5112	-1.37	0.195
ATTITUDE	1.2155	0.1895	6.41	0.000

S = 0.3930   R-SQ = 76.0%   R-SQ(ADJ) = 74.1%

ANALYSIS OF VARIANCE

SOURCE	DF	SS	MS	F	P
REGRESSION	1	6.3522	6.3522	41.12	0.000
ERROR	13	2.0080	0.1545		
TOTAL	14	8.3602			

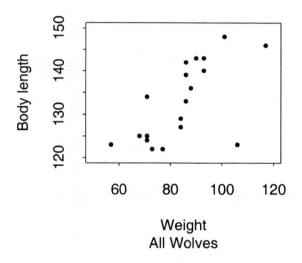

Weight
All Wolves

11.43 Using MINITAB, and removing case 18, we obtain the output

Regression Analysis

```
The regression equation is
bodylengr = 88.7 + 0.538 weightr

Predictor        Coef       StDev        T       P
Constant       88.681       7.648      11.60   0.000
weightr        0.53793     0.09084      5.92   0.000


S = 5.180        R-Sq = 68.7%

Analysis of Variance

Source          DF         SS          MS        F       P
Regression       1       940.96      940.96    35.07   0.000
Residual Error  16       429.32       26.83
Total           17      1370.28
```

(a) The fitted equation is $\hat{y} = 88.68 + .5379x$.

(b) We first calculate $S_{xx} = \sum x_i^2 - (\sum x_i)^2/n = 127586 - (1496)^2/18 = 3251.8$

With $\alpha = .05$, we test $H_0 : \beta_1 = 0$ versus $H_1 : \beta_1 > 0$. The rejection region is $R : T > t_{.05} = 1.746$ for 16 degrees of freedom. Here

$$t = \frac{\hat{\beta}_1}{s/\sqrt{S_{xx}}} = \frac{.5379}{5.180/\sqrt{3251.8}} = 5.92$$

or, according to the output $t = .5379/.09084 = 5.92$. Consequently, we reject $H_0 : \beta_1 = 0$ in favor of $H_1 : \beta_1 > 0$. The P-value is less than .001.

(c) The estimated slope has increased by over 30 %. The test for slope equal 0 has an even smaller P-value. Also the estimate of variance, s, is smaller with case 18 removed.

11.45 $r^2 = \dfrac{(9.6)^2}{(14.2)(18.3)} = .355$

11.47 By Exercise 11.24, we have $S_{xx} = 26$, $S_{yy} = 10$ and $S_{xy} = -16$.

(a) $r^2 = \dfrac{(-16)^2}{(26)(10)} = .985$

(b) $r = \dfrac{-16}{\sqrt{26}\sqrt{10}} = -.992$

(c) $SSE = 10 - \dfrac{(-16)^2}{26} = .1538$

(d) $s^2 = \dfrac{.1538}{5-2} = .0513$

11.49 Proportion explained $= r^2 = .302$.

11.51 (a) $r = .649$ and $r^2 = .421$

(b) $r = .279$ and $r^2 = .078$

(c) $r = .733$ and $r^2 = .537$

(d) The pattern is quite different for male and female wolves.

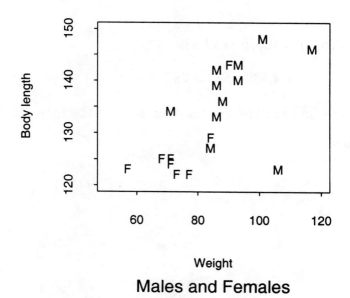

Males and Females

11.53 SS due to regression $= \dfrac{S_{xy}^2}{S_{xx}}$

$$= \frac{S_{xy}^2}{S_{xx}} \frac{S_{xx}}{S_{xx}} \quad (\text{multiply by } \tfrac{S_{xx}}{S_{xx}})$$

$$= \hat{\beta}_1^2 S_{xx}.$$

11.55 (b) $\bar{x} = 25/9 = 2.778$, $\bar{y} = 125/9 = 13.889$

$S_{xx} = 21.556$, $S_{yy} = 304.889$, $S_{xy} = 78.778$

(a) & (c) $\hat{y} = 3.737 + 3.655x$. The scatter diagram and the fitted line are given in Figure 11.6.

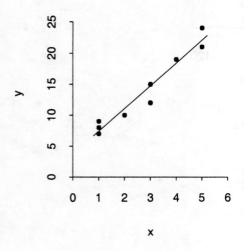

Figure 11.6: Exercise 11.55. Scatter diagram and the fitted line.

(d) predicted value $= 3.737 + 3.655(3) = 14.70$.

11.57 (a) $\hat{\beta}_1 \pm t_{.025} \dfrac{s}{\sqrt{S_{xx}}} = 3.655 \pm 2.365 \dfrac{1.558}{\sqrt{21.556}} = 3.655 \pm .794$

or $(2.86, 4.45)$.

(b) $3.737 + 3.655(4) \pm 1.895(1.558)\sqrt{\dfrac{1}{9} + \dfrac{(4 - 2.778)^2}{21.556}}$

or $(17.10, 19.61)$.

11.59 (a) The decomposition of the total y-variability is

$$S_{yy} \quad = \quad \dfrac{S_{xy}^2}{S_{xx}} \quad + \quad SSE$$

Total	Explained by linear relation	Error
38.7 =	27.46	+ 11.24

(b) proportion explained $= r^2 = \dfrac{27.46}{38.7} = .71$.

(c) $r = \dfrac{-12.4}{\sqrt{5.6}\sqrt{38.7}} = -.84$.

11.61 (a) $r = S_{xy}/\sqrt{S_{xx}S_{yy}} = 43215/\sqrt{35638 \times 58121.88} = .9495$

(b) Proportion of variability explained $= r^2 = .902$ and this is quite high.

11.63 (a) Predicted value $= 9.81 - .868(5) = 5.47$.

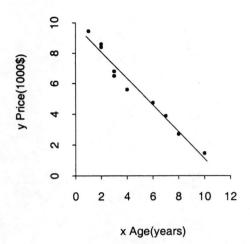

A 95% confidence interval is calculated as

$$5.47 \pm 2.306(.505)\sqrt{\tfrac{1}{10} + \frac{(5-4.6)^2}{80.4}} \quad \text{or} \quad (5.10, \ 5.84)$$

(b) Predicted value for a single car = 5.47.

A 90% confidence interval is calculated as

$$5.47 \pm 1.860(.505)\sqrt{1 + \tfrac{1}{10} + \frac{(5-4.6)^2}{80.4}} \quad \text{or} \quad (4.48, \ 6.46)$$

(c) No. There is no data over that region. The linear relation cannot hold for cars that old because we know the selling price never goes below zero.

11.65 (a) $S_{xy} = 21 - (17)(31)/20 = -5.35$

$S_{xx} = 19 - (17)^2/20 = 4.55$

$S_{yy} = 73 - (31)^2/20 = 24.95$

$\hat{\beta}_1 = \frac{-5.35}{4.55} = -1.1758, \hat{\beta}_0 = 1.55 + (1.1758)(0.85) = 2.549, \hat{y} = 2.55 - 1.176x.$

(b) $r = \dfrac{-5.35}{\sqrt{(4.45)(24.95)}} = -.502$

(c) $r^2 = .252$ so the proportion of y-variability explained by the straight line fit is quite low.

11.67 The MINITAB output is

```
Regression Analysis


The regression equation is
weight = - 53.2 + 1.03 bodyleng

Predictor       Coef        StDev          T         P
Constant       -53.17       39.21      -1.36     0.193
bodyleng       1.0349       0.2945      3.51     0.003

S = 11.30      R-Sq = 42.1%      R-Sq(adj) = 38.7%

Analysis of Variance

Source          DF         SS         MS        F        P
Regression       1      1577.2     1577.2    12.35   0.003
Residual Error  17      2170.9      127.7
Total           18      3748.1
```

```
Unusual Observations
Obs bodyleng  weight     Fit StDev Fit Residual St Resid
 18        123  106.00  74.13      3.89     31.87     3.00R
```

R denotes an observation with a large standardized residual

(a) The scatter plot is

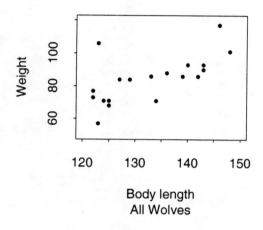

(b) From the output,

the least squares regression line is $\hat{y} = -2.012 + 1.396x$. $\hat{y} = -53.17 +$ $1.0349x$. According to the small $P-$value for the test of zero intercept, the model could be re-fit without an intercept term. However, the unusual observation noted in the output could make the current analysis misleading(see Exercise 11.68). We proceed as if the intercept term is needed.

(c)

With d.f. $= 17$, we have $t_{.05} = 1.740$. Therefore, the null hypothesis $H_0 : \beta_1 = 0$ will be rejected with $\alpha = .05$ if $T \geq 1.740$. According to the output

$$T = \frac{1.0349}{0.2945} = 3.51$$

Consequently, we reject $H_0 : \beta_1 = 0$ in favor of the alternative hypothesis

$H_1: \beta_1 > 0$. Since the $P-$value$= 0.003$ is very small, the data strongly support that $\beta_1 > 0$ which, in turn, indicates that the expected value of weight increases with body length.

(d) The answers are different although both slopes are positive. Here we minimized the sum of squares concerning weight and in Exercise 11.42 we minimized the sum of squares concerning body length.

11.69 (a) From the data, we calculate

$$\bar{x} = 154.7,\ \bar{y} = 616.1,\ S_{xx} = 11262.2,\ S_{yy} = 225897.8,\ S_{xy} = 41914.6,$$

$$\hat{\beta}_1 = 3.7217,\ \hat{\beta}_0 = 40.35.$$

The fitted line is $\hat{y} = 40.35 + 3.722x$.

(b) The residual sum of squares is $SSE = 69904$ and the estimate of σ is

$s = \sqrt{SSE/18} = 62.3$. Since $t_{.025} = 2.101$, for 18 degrees of freedom,

$$\hat{\beta}_1 \pm t_{.025}\frac{s}{\sqrt{S_{xx}}} = 3.722 \pm 2.101\frac{62.3}{\sqrt{11262.2}}\quad \text{or}\quad (2.489,\ 4.955)$$

(c) The predicted value of y for $x^* = 150$ is

$$40.35 + 3.722 \times 150 \approx 599.$$

To obtain the 95% confidence interval for a single CLEP score at CQT = 150, we calculate

$$t_{.025}\ s\sqrt{1 + \frac{1}{n} + \frac{(x^* - \bar{x})^2}{S_{xx}}} = 2.101 \times 62.3\ \sqrt{1 + \frac{1}{20} + \frac{(150 - 154.7)^2}{11262.2}}$$

or ≈ 134.

The required confidence interval is 599 ± 134 or $(465,\ 733)$.

(d) The calculations are similar to those in part (c) and we present the final results only:

At $x = 175$, the confidence interval is 692 ± 136 or $(556,\ 828)$.

At $x = 195$, the confidence interval is 766 ± 143 or $(623, 909)$.

11.71 Using MINITAB, the relevant data are in DBT6.DAT.

```
The regression equation is
ozone = 1952 - 0.810 year

Predictor        Coef        Stdev      t-ratio          p
Constant        1951.7       293.6         6.65      0.000
year           -0.8104       0.1484       -5.46      0.000

s = 7.750        R-sq = 49.9%      R-sq(adj) = 48.2%

ANALYSIS OF VARIANCE

SOURCE         DF          SS          MS         F       p
Regression      1      1791.7      1791.7     29.83   0.000
Error          30      1801.8        60.1
Total          31      3593.5
```

We want to test hypotheses $H_0 : \beta_1 = 0$ versus $H_1 : \beta_1 < 0$. With $\alpha = .05$, the rejection region is $R : T < t_{.05} = -1.697$ for 30 degrees of freedom. Since the t statistic is -5.46, the null hypothesis is rejected in favor of $\beta_1 < 0$. Since the P-value is less than $.001$, there is very strong evidence that the total ozone in Toronto is decreasing over the years.

Chapter 12

Regression Analysis - II Multiple Linear Regression And Other Topics

12.1 (a) The scatter diagram is

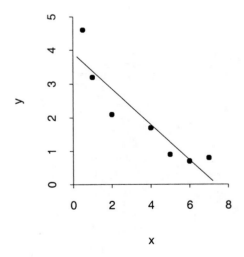

Figure 12.1: Exercise 2.1. Scatter diagram and the fitted line

(b) $\bar{x} = \frac{25.5}{7} = 3.6429$, $\bar{y} = \frac{14.0}{7} = 2.0$,

$S_{xx} = 38.3571$, $S_{xy} = -20.2$, $S_{yy} = 12.64$

$$\hat{\beta}_1 = \frac{S_{xy}}{S_{xx}} = -.5266$$

$$\hat{\beta}_0 = \bar{y} - \hat{\beta}_1 \bar{x} = 2.0 - (-.5266)(3.6429) = 3.9184$$

The fitted line $\hat{y} = 3.92 - .53x$ is shown in Figure 12.1.

(c) Proportion of y variability explained $= r^2 = \dfrac{S_{xy}^2}{S_{xx}S_{yy}} = .842.$

This is relatively high but the scatter diagram reveals a relation on a curve.

12.3 (a) $y' = \dfrac{1}{y^{1/3}}$, $x' = x$

(b) $y' = \dfrac{1}{y}$, $x' = \dfrac{1}{1+x}$

12.5   (a) The scatter diagram is shown in Figure 12.3(i).

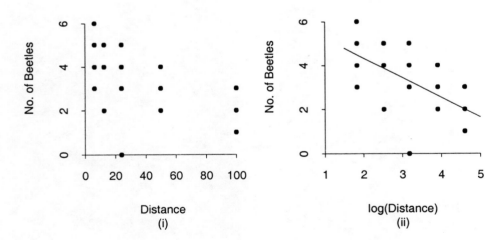

Figure 12.3: Exercise 12.5. Scatter diagrams and the fitted line

(b) The scatter diagram of the transformed data, $x' = \log x$ and $y' = \log y$, reveals a more nearly linear relationship as shown in Figure 12.3(ii). Using MINITAB, the original data are set in columns C1 and C2.

```
        NAME C1 'DISTANCE' C2 'BEETLES' C3 'LOGEDIST'
        LOGE C1 SET IN C3
        REGRESS Y IN C2 ON 1 PREDICTOR IN C3
```

THE REGRESSION EQUATION IS
BEETLES = 6.14 - 0.899 LOGEDIST

```
PREDICTOR        COEF        STDEV     T-RATIO        P
CONSTANT        6.1e74      0.9918        6.19     0.000
LOGEDIST       -0.8993      0.2954       -3.04     0.007
```

S = 1.295   R-SQ = 34.0%   R-SQ(ADJ) = 30.3%

ANALYSIS OF VARIANCE

```
SOURCE          DF          SS          MS          F        P
REGRESSION       1      15.547      15.547       9.27    0.007
ERROR           18      30.203       1.678
TOTAL           19      45.750
```

In order to evaluate the mean and sum of squares for log(x) we also calculate:

```
            STDEV C3
ST.DEV.  =       1.0059
            MEAN C3
MEAN     =       3.2107
```

(c) Since $S_{xx} = (n-1)(\text{ST.DEV.})^2 = 19(1.0059)^2 = 19.225$ and $t_{.025} = 2.101$ for 18 d.f., a 95% confidence interval for β_1 is calculated as

$$\hat{\beta}_1 \pm t_{.025}\frac{s}{\sqrt{S_{xx}}} = -.8993 \pm (2.101)(.2954) \quad \text{or } (-1.52, \ -.28).$$

(d) At $x = 18$, $\hat{y} = 6.14 - .899\log_e(18) = 3.54$. Since $S_{xx} = 19.225$, a 95% confidence interval is calculated as

$$3.54 \pm 2.101(1.295)\sqrt{\frac{1}{20} + \frac{(2.8904 - 3.2107)^2}{19.225}} = 3.54 \pm .64$$

or $(2.90, 4.18)$.

12.7   (a) & (b) In Figure 12.4(i), the scatter diagram shows a relationship, along a
curve, between the diameters and the heights of sugar maple trees. Using
transformations $x' = \log x$ and $y' = \log y$, the scatter diagram of the

transformed data reveals a linear relationship as shown in Figure 12.4(ii).

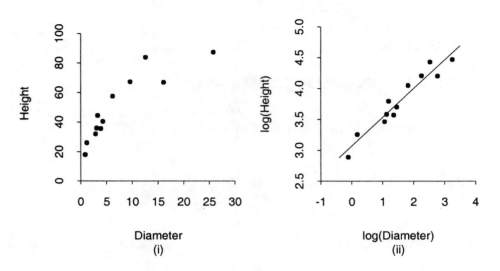

Figure 12.4: Exercise 12.7. Scatter diagrams and the fitted line

(c) Using MINITAB, the original data are set in columns C1 and C2. The fitted line, $\widehat{\log_e(\text{height})} = 3.07 + .465 \log_e(\text{diameter})$, is shown in Figure 12.4(ii).

```
LOGE C1 SET IN C3
LOGE C2 SET IN C4
NAME C3 'LOGDIAM' C4 'LOGHEIG'
REGRESS Y IN C4 ON 1 PREDICTOR IN C3

THE REGRESSION EQUATION IS
LOGHEIG = 3.07 + 0.465 LOGDIAM

PREDICTOR       COEF       STDEV     T-RATIO        P
CONSTANT      3.06980     0.07467      41.11    0.000
LOGDIAM       0.46459     0.04039      11.50    0.000

S = 0.1348     R-SQ = 93.0%     R-SQ(ADJ) = 92.3%

ANALYSIS OF VARIANCE

SOURCE          DF          SS          MS          F          P
```

```
        REGRESSION    1      2.4031      2.4031    132.30    0.000
        ERROR        10      0.1816      0.0182
        TOTAL        11      2.5848

            CORR C3 C4

        CORRELATION OF LOGDIAM AND LOGHEIG = 0.964
```

(d) Proportion of variance of $\log_e y$ explained $= r^2 = .930$.

12.9 When $x_1 = 3$ and $x_2 = -2$, the mean of the response Y is

$$E(Y) = \beta_0 + \beta_1 x_1 + \beta_2 x_2 = -2 - 1(3) + 3(-2) = -11$$

12.11 (a) Since $t_{.025} = 2.110$ for 17 d.f., 95% confidence intervals for β_0 and β_2 are

$$\hat{\beta}_0 \pm t_{.025} \times S.E.(\hat{\beta}_0) = 3.8 \pm 2.110(1.345) \quad \text{or} \quad (.96,\ 6.64)$$

$$\hat{\beta}_2 \pm t_{.025} \times S.E.(\hat{\beta}_2) = -4.2 \pm 2.110(.4056) \quad \text{or} \quad (-5.06,\ -3.34)$$

(b) Given $\alpha = .05$, the rejection region is $R : T > t_{.05} = 1.740$ for 17 degrees of freedom. Since

$$\frac{\hat{\beta}_1 - 6}{S.E.(\hat{\beta}_1)} = \frac{7.9 - 6}{.907} = 2.09$$

we reject the hypothesis $H_0 : \beta_1 = 6$ in favor of $H_1 : \beta_1 > 6$ at level $\alpha = .05$.

12.13   (a) Using MINITAB we obtain

```
The regression equation is
McannineL = 24.8 + 1.41 Mage


Predictor      Coef     StDev        T        P
Constant    24.7755    0.8693    28.50    0.000
Mage         1.4127    0.2958     4.78    0.001


S = 1.274     R-Sq = 71.7%  R-Sq(adj) = 68.6%

Analysis of Variance

Source          DF      SS        MS       F      P
Regression       1   37.014    37.014   22.81  0.001
Residual Error   9   14.606     1.623
Total           10   51.620
```

(b) The least squares fit to a quadratic function reduces the residual sum of
squares from 14.606 to 12.7044 but the $P-$ value for testing $H_0 : \beta$
is large so the quadratic term in not needed.

```
Polynomial Regression


Y = 23.9111 + 2.57778X - 0.255556X**2
R-Sq = 75.4 %
```

```
Analysis of Variance

SOURCE        DF      SS       MS       F         P
Regression     2   38.9156  19.4578  12.2526  3.67E-03
Error          8   12.7044   1.5881
Total         10   51.6200


SOURCE        DF    Seq SS       F        P
Linear         1   37.0139  22.8073  1.01E-03
Quadratic      1    1.9016   1.19746  0.305680
```

(c) The proportion of $y-$ variability explained is $R^2 = .754$ or 75.4 %.

(d) For the straight line regression $s = 1.274$ and for the quadratic fit is $s = \sqrt{\text{Error MS}} = \sqrt{1.5881} = 1.260$.

12.15 (a) $\hat{\beta}_0 = -.0810, \quad \hat{\beta}_1 = .64588, \quad$ and $\quad \hat{\beta}_2 = .8046$

   (b) $\hat{y} = -.081 + .646x_1 + .805x_2$. The constant term could be dropped and the model refit.

   (c) Proportion of the y-variability explained $= R^2 = .865$

   (d) $s^2 = \dfrac{SSE}{n-2} =$ Error MS $= .01023$

12.17 (a) Since $R : |T| > t_{.025} = 2.074$ with 22 d.f. and the t-ratio for $\hat{\beta}_1$ is 7.02, we reject $H_0 : \beta_1 = 0$ in favor of $H_1 : \beta_1 \neq 0$ at level $\alpha = .05$.

   (b) Since $R : |T| > t_{.025} = 2.074$ with 22 d.f. and the t-ratio for $\hat{\beta}_2$ is 3.29, we reject $H_0 : \beta_2 = 0$ in favor of $H_1 : \beta_2 \neq 0$ at level $\alpha = .05$.

   (c) $\hat{y} = -.081 + .646(1.9) + .805(1.0) = 1.951$

   (c) A 90% confidence interval for β_0, which includes 0, is

$$\hat{\beta}_0 \pm t_{.05} \times S.E.(\hat{\beta}_0) = -.081 \pm 1.717(.1652) \quad \text{or} \quad (-.203, \ .365)$$

12.19 Using MINITAB, the initial run time, the final run time and the final row time are set in columns C4, C5 and C7 respectively. The fitted equation is

final run time = 95.14 + .6324(initial run time) + .13594(final row time)

For $\alpha = .079$, final row time could be dropped from the model. The re-fitted linear model is already given in Exercise 11.72.

```
        NAME C4 'PRE_RUN' C5 'POST_RUN' C7 'POST_ROW'
        REGRESS Y IN C5 ON 2 PREDICTORS IN C4 C7


THE REGRESSION EQUATION IS
POST_RUN = 95.1 + 0.632 PRE_RUN + 0.136 POST_ROW

PREDICTOR        COEF      STDEV   T-RATIO        P
CONSTANT        95.14      39.43      2.41    0.018
PRE_RUN       0.63240    0.04879     12.96    0.000
POST_ROW      0.13594    0.07629      1.78    0.079


S = 43.10     R-SQ = 81.6%   R-SQ(ADJ) = 81.2%


ANALYSIS OF VARIANCE
SOURCE        DF       SS        MS        F       P
REGRESSION     2   644306    322153   173.44   0.000
ERROR         78   144877      1857
TOTAL         80            789183


SOURCE          DF      SEQ SS
PRE_RUN          1      638408
POST_ROW         1        5898
```

12.21 (a) The scatter diagram of y versus $x' = \log_{10} x$ is shown in Figure 12.5.

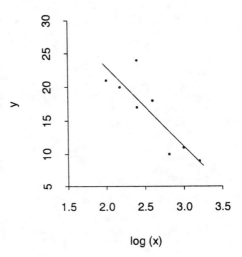

Figure 12.5: Exercise 12.21. Scatter diagram and the fitted line

(b) The MINITAB output is given below. The fitted line

$$\hat{y} = 46.55 - 11.77 \log_{10}(x)$$

is shown in Figure 12.5.

```
The regression equation is
y = 46.6 - 11.8 logt(x)

Predictor      Coef     StDev        T        P
Constant     46.550     7.242     6.43    0.001
logt(x)     -11.772     2.783    -4.23    0.005
```

```
S = 3.031        R-Sq = 74.9%

Analysis of Variance

Source          DF      SS       MS      F      P
Regression       1    164.39   164.39  17.90  0.005
Residual Error   6     55.11     9.19
Total            7    219.50
```

(c) By the MINITAB output in part (a), $\dfrac{s}{\sqrt{S_{x'x'}}} = 2.783$. You could also get the same result by direct calculation using $S_{x'x'} = 1.1862$. A 90% confidence interval for β_1 is calculated as

$$\hat{\beta}_1 \pm t_{.05}\frac{s}{\sqrt{S_{x'x'}}} = -11.77 \pm 1.943(3.031) \quad \text{or} \quad (-17.66,\ -5.88).$$

(d) At $x = 300$, $\hat{y} = 46.55 - 11.772(\log_{10}(300)) = 17.39$. Since $\overline{x'} = 2.5739$ and $S_{x'x'} = 1.1862$, a 95% confidence interval for the expected y-value at $x = 300$ is

$$\hat{y} \pm t_{.025}s\sqrt{\frac{1}{n} + \frac{((x')^* - \overline{x'})^2}{S_{x'x'}}}$$

$$= 17.39 \pm 2.447(3.031)\sqrt{\frac{1}{8} + \frac{(2.4771 - 2.5739)^2}{1.1862}} = 17.39 \pm 2.72$$

or $(14.7, 20.1)$.

12.23 (a) & (b) The scatter diagram of y versus x shown in Figure 12.6(i) reveals a relation along a curve and that of $y' = \log_{10} y$ versus x shown in Figure 12.6(ii) looks like a straight line relation.

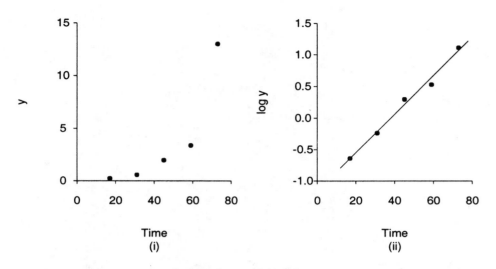

Figure 12.6: Exercise 12.23. Scatter diagrams and the fitted line

(c) The fitted line $\hat{\log}_{10} y = -1.16 + .0305x$ is shown in Figure 12.6(ii).

The MINITAB output is given below.

```
NAME C1 'TIME' C2 'Y' C3 'LOG(Y)'
LOGT C2 SET C3
REGRESS Y IN C3 ON 1 PREDICTOR IN C1

THE REGRESSION EQUATION IS
LOG(Y) = - 1.16 + 0.0305 TIME

PREDICTOR      COEF      STDEV    T-RATIO        P
CONSTANT    -1.15984   0.09468    -12.25    0.001
TIME        0.030513  0.001926     15.84    0.001

S = 0.08526    R-SQ = 98.8%    R-SQ(ADJ) = 98.4%

ANALYSIS OF VARIANCE

SOURCE      DF      SS       MS        F        P
REGRESSION   1   1.8249   1.8249   251.05    0.001
ERROR        3   0.0218   0.0073
TOTAL        4   1.8467
```

12.25 (a) A 90% confidence interval for β_1 is calculated as

$$\hat{\beta}_1 \pm t_{.05} \times S.E.(\hat{\beta}_1) = 1.93 \pm 1.717(.062) \quad \text{or} \quad (1.82,\ 2.04)$$

(b) Given $\alpha = .05$, the rejection region is $R : T < t_{.05} = -1.717$ for 22 degrees of freedom. Since

$$\frac{\hat{\beta}_2 - 25}{S.E.(\hat{\beta}_2)} = \frac{20.2 - 25}{2.43} = -1.975$$

we reject $H_0 : \beta_2 = 25$ in favor of $H_1 : \beta_2 < 25$ at the 5% significance level.

12.27   (a) $\hat{y} = 50.4 + .1907x_2$ and $r^2 = .03$

```
              NAME C1 'X1' C2 'X2' C3 'X3' C4 'Y'
              REGRESS Y IN C4 ON 1 PREDICTOR IN C2

        THE REGRESSION EQUATION IS
        Y = 50.4 + 0.191 X2

        PREDICTOR    COEF    STDEV   T-RATIO       P
        CONSTANT     50.40   17.57    2.87      0.010
        X2           0.1907  0.2561   0.74      0.466

        S = 19.95        R-SQ = 3.0%  R-SQ(ADJ) = 0.0%

        ANALYSIS OF VARIANCE

        SOURCE       DF      SS       MS      F      P
        REGRESSION   1     220.8    220.8   0.55   0.466
        ERROR        18   7164.2    398.0
        TOTAL        19   7385.0
```

(b) $\hat{y} = -92.32 + .583x_1 - .1494x_2 + 35.07x_3$ and $R^2 = .586$

```
           REGRESS Y IN C4 ON 3 PREDICTORS IN C1 C2 C3

        THE REGRESSION EQUATION IS
        Y = - 92.3 + 0.583 X1 - 0.149 X2 + 35.1 X3

        PREDICTOR    COEF    STDEV   T-RATIO       P
        CONSTANT    -92.32   33.22   -2.78      0.013
        X1           0.5830  0.3206   1.82      0.088
        X2          -0.1494  0.1923  -0.78      0.449
        X3          35.07   11.25     3.12      0.007
```

```
S = 13.83    R-SQ = 58.6%     R-SQ(ADJ) = 50.8%

ANALYSIS OF VARIANCE

SOURCE       DF     SS      MS     F     P
REGRESSION    3  4325.7  1441.9  7.54 0.002
ERROR        16  3059.2   191.2
TOTAL        19  7385.0
SOURCE       DF     SEQ SS
X1            1     2466.6
X2            1        1.4
X3            1     1857.8
```

(c) Even three variables do not predict well. In fact, the GPA (x_3) could predict almost as well by itself. We summarize the results in the following table.

Predictor	x_3	x_3 and x_1	x_3, x_1 and x_2
R^2	.495	.570	.586

12.29 The design matrix $\mathbf{X}$ of the model $Y = \beta_0 + \beta_1 x_1 + \beta_2 x_2 + e$ is

$$\mathbf{X} = \begin{bmatrix} 1 & 1 & 8.1 \\ 1 & 2 & 17.0 \\ 1 & 2 & 12.6 \\ 1 & 3 & 18.4 \\ 1 & 3 & 19.5 \\ 1 & 4 & 29.2 \\ 1 & 6 & 40.4 \\ 1 & 7 & 51.6 \\ 1 & 8 & 62.6 \\ 1 & 10 & 80.1 \end{bmatrix}$$

12.31 We plot the residual versus the predicted value $\hat{y}$ in Figure 12.7. The plot shows that the residuals tend to have higher variability with increasing values of $\hat{y}$. This indicates a violation of the assumption of constant variances.

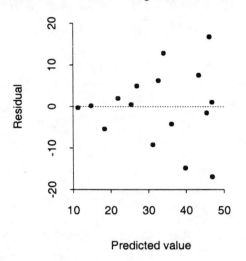

Figure 12.7: Plot of residual vs. predicted value for Exercise 12.31.

12.33 (a) We plot the residual versus the predicted value $\hat{y}$ and the time order respectively in Figure 12.8.

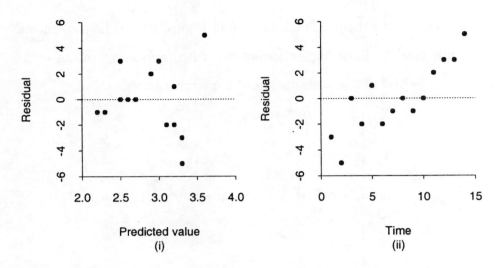

Figure 12.8: Residual vs. predicted value and time order for Exercise 12.33.

(b) The plot (i) does not seem to signal any appreciable violation of the assumptions. The plot (ii) of residual versus time order, however, shows a distinct pattern. The residuals tend to steadily increase in time. This indicates a possible violation of the independence assumption.

12.35 Looking at the residuals in time order shown in Figure 12.9, a distinct pattern is visible. The residuals decrease quite systematically till about the year 15 and then they steadily increase. Residuals adjacent in time have similar magnitude. This pattern casts serious doubt on independence assumption. Violation of independence is frequent in time series data such as these.

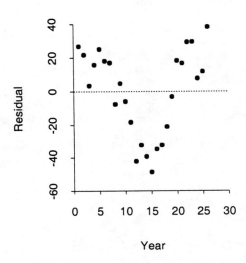

Figure 12.9: Plot of residual vs. time order for Exercise 12.35.

Chapter 13

ANALYSIS OF CATEGORICAL DATA

13.1 The null hypothesis is $H_0 : p_1 = p_2 = \cdots = p_6 = \frac{1}{6}$. Under H_0, the expected frequency of each cell is $320 \times \frac{1}{6}$. The χ^2 statistic for goodness-of-fit is calculated as follows:

Face no.	1	2	3	4	5	6	Total
Observed frequency (O)	39	63	56	67	57	38	320
Expected frequency (E)	$\frac{320}{6}$	$\frac{320}{6}$	$\frac{320}{6}$	$\frac{320}{6}$	$\frac{310}{6}$	$\frac{320}{6}$	320
$\frac{(O-E)^2}{E}$	3.852	1.752	0.133	3.502	0.252	4.408	$13.90 = \chi^2$

We take $\alpha = .05$. For d.f. $= 5$ we find $\chi^2_{.05} = 11.07$ so the rejection region is $R : \chi^2 \geq 11.07$. Since the observed χ^2 is larger than $\chi^2_{.05}$, the null hypothesis is rejected at $\alpha = .05$. The model of a fair die is contradicted.

13.3 The null hypothesis is $H_0 : p_1 = .4, p_2 = .4, p_3 = .1, p_4 = .1$. Multiplying these probabilites by $n = 100$, the expected cell frequencies are found to be 40, 40, 10, and 10, respectively.

Blood type	0	A	B	AB	Total
O	40	44	10	6	100
E	40	40	10	10	100
$\frac{(O-E)^2}{E}$	0.00	0.40	0.00	1.60	$2.00 = \chi^2$
					d.f. $= 3$

With $\alpha = .05$ and d.f. $= 3$, we find $\chi^2_{.05} = 7.81$ so the rejection region is $R : \chi^2 \geq 7.81$. The observed $\chi^2 = 1.64$ is not in R. Therefore, H_0 is not rejected at $\alpha = .05$.

13.5 Letting p_1, p_2, p_3, and p_4 denote the true proportions of walnuts, hazelnuts, almonds, and pistachios, respectively, we are to test the null hypothesis

$$H_0 : \ p_1 = .45, p_2 = .20, p_3 = .20, p_4 = .15.$$

The expected frequencies and the χ^2 statistic are calculated below.

	Walnuts	Hazelnuts	Almonds	Pistachios	Total
O	95	70	33	42	240
E	108	48	48	36	240
$\dfrac{(O-E)^2}{E}$	1.565	10.083	4.688	1.000	$17.336 = \chi^2$
					d.f. $= 3$

With $\alpha = .025$ and d.f. $= 3$, we find $\chi^2_{.025} = 9.35$ so the rejection region is $R : \ \chi^2 \geq 9.35$. Since the observed χ^2 is in R, the null hypothesis is rejected at $\alpha = .025$. We conclude that there is strong evidence of mislabeling.

13.7 The null hypothesis that all 7 days of the week are equally likely is formalized as

$$H_0 : \ p_1 = p_2 = \cdots = p_7 = \frac{1}{7}.$$

Since the observed frequencies are given in units of 10,000, we proceed with the calculation of the expected frequencies and the χ^2 statistic in the same units. In particular, the expected frequency of each cell is $356.31 \times \frac{1}{7} = 50.901$.

	Mon	Tues	Wed	Thurs	Fri	Sat	Sun	Total
O	52.09	54.46	52.68	51.68	53.83	47.21	44.36	356.31
E	50.901	50.901	50.901	50.901	50.901	50.901	50.901	356.31
$\frac{(O-E)^2}{E}$	.028	.249	.062	.012	.169	.268	.841	1.629

The observed $\chi^2 = 1.629 \times 10,000 = 16290$. Taking $\alpha = .01$ we find that $\chi^2_{.01} = 16.81$ with d.f. $= 6$. Since the observed χ^2 is in the rejection region $R : \chi^2 \geq 16.81$, H_0 is rejected at $\alpha = .01$. The Sunday count is low.

(Remark: These calculations illustrate the fact that with a very large sample size, seemingly small deviations of the relative frequencies from the probabilities specified by a model turn out to be statistically significant.)

13.9 (a) From the binomial table for $n = 3$ and $p = .4$ we find

$$P[X = 0] = \ .216$$

$$P[X = 1] = \ .648 - .216 = .432$$

$$P[X = 2] = \ .936 - .648 = .288$$

$$P[X = 3] = 1.000 - .936 = .064$$

(b) Let p_0, p_1, p_2, and p_3 denote the probabilities of the four categories: 0, 1, 2, and 3 males in litter, respectively, The null hypothesis is

$$H_0: \ p_0 = .216, p_1 = .432, p_2 = .288, p_3 = .064.$$

The calculations for the χ^2 test are given in the following table:

No. of males	0	1	2	3	Total
Observed frequency (O)	19	32	22	7	80
Expected frequency (E)	17.28	34.56	23.04	5.12	80
$\dfrac{(O-E)^2}{E}$	.171	.190	.047	.690	$1.098 = \chi^2$
					d.f. $= 2$

Here d.f. $= \#$ cells $- 1 - \#$ parameters estimated

$$= 4 - 1 - 1 = 2.$$

We take $\alpha = .05$, and find that $\chi^2_{.05} = 5.99$ with d.f. $= 2$. Since the observed χ^2 is less than $\chi^2_{.05}$, the null hypothesis is not rejected at $\alpha = .05$. We conclude that the binomial model is not contradicted.

13.11 Let p_1 and p_1 be the probabilities that at person opens all their mail for males and females, respectively.

(a)

	Open all mail	Don't open all mail	Total
Males	414	386	800
Females	532	368	900
Total	946	754	1700

(b) Let p_1 and p_2 be the probabilities that at person opens all their mail for males and females, respectively.

(b) The null hypothesis is $H_0 : p_1 = p_2$.

(c) With $\alpha = .05$ and d.f. $= 1$, we find $\chi^2_{.05} = 3.84$ so the rejection region is $R : \chi^2 \geq 3.84$. The χ^2 statistic for testing homogeneity is calculated as follows:

Expected Values

	Open all mail	Don't open all mail	Total
Males	445.176	354.824	800
Females	500.823	399.176	900
Total	946	754	1700

$(O - E)^2/E$

	Open all mail	Don't open all mail
Males	2.183	2.739
Females	1.941	2.435
Total	946	754

Chi-Sq $= 2.183 + 2.739 + 1.941 + 2.435 = 9.298$ so we reject the null hypothesis, of equal probabilities of openning all mail, at the 5 % level of significance. The P−value is about .02. The largest contribution to χ^2

comes from the mail-don't open cell where the observed count is higher than expected. The female-don't open count is lower than expected.

13.13 Denote by p_{A1}, p_{A2}, p_{A3}, and p_{A4} the probabilities of response in the categories 'none', 'slight', 'moderate', and 'severe', respectively, under the use of Brand A pills, and similarly, $p_{B1}, \ldots, p_{B4}$ for Brand B. We are to test

$$H_0: \ p_{Aj} = p_{Bj}, j = 1, 2, 3, 4.$$

The χ^2 statistic for testing homogeneity is calculated as follows:

```
Expected counts are printed below observed counts

          C1       C2       C3       C4     Total
  1       18       17        6        4        45
        14.50    15.50    10.00     5.00

  2       11       14       14        6        45
        14.50    15.50    10.00     5.00

Total     29       31       20       10        90

ChiSq =  0.845 +  0.145 +  1.600 +  0.200 +
         0.845 +  0.145 +  1.600 +  0.200 = 5.580
df = 3
```

With $\alpha = .05$ and d.f. $= 3$, we find $\chi^2_{.05} = 7.81$. Because the observed χ^2 is smaller than $\chi^2_{.05}$, H_0 is not rejected at $\alpha = .05$. We fail to conclude that the two pills are significantly different in quality.

13.15   (a) Denote by p_1, p_2, p_3, and p_4 the probabilities that a paper will be taken from sites 1, 2, 3, and 4 We are to test $H_0 : p_1 = p_2 = p_3 = p_4$.

(b) First we must calculate the number of papers taken from each site. For site 1, the number taken is $50 - 17 = 33$. The χ^2 statistic for testing homogeneity is calculated as follows:

```
Chi-Square Test


Expected counts are printed below observed counts

        Remain    Taken    Total
   1      17        33        50
        14.62     35.38

   2      12        35        47
        13.74     33.26

   3       7        41        48
        14.03     33.97

   4      21        29        50
        14.62     35.38

Total     57       138       195

   Chi-Sq =  0.389 +  0.161 +
             0.220 +  0.091 +
             3.523 +  1.455 +
             2.789 +  1.152 = 9.780
   DF = 3, P-Value = 0.021
```

With $\alpha = .05$ and d.f. $= 3$, we find $\chi^2_{.05} = 7.81$ so the rejection region is $R : \chi^2 \geq 7.81$. Since the observed $\chi^2 = 9.780$ is in R, we reject H_0 at $\alpha = .05$ and conclude that the proportions of papers taken differ between the sites.

(c) A 95% confidence interval for a population proportion p is given by

$$\hat{p} \pm 1.96\sqrt{\hat{p}\hat{q}/n}$$

We calculate

$$1: \quad \hat{p}_1 = \frac{33}{50} = .66, \quad .66 \pm 1.96\sqrt{\frac{.66 \times .34}{50}} = .66 \pm .13$$

$$\text{or} \quad (.53, .79)$$

$$2: \quad \hat{p}_2 = \frac{35}{47} = .745, \quad .475 \pm .125 \quad \text{or} \quad (.35, .60)$$

$$3: \quad \hat{p}_3 = \frac{41}{48} = .854, \quad .854 \pm .099 \quad \text{or} \quad (.76, .95)$$

$$4: \quad \hat{p}_4 = \frac{29}{50} = .58, \quad .58 \pm .137 \quad \text{or} \quad (.44, .62)$$

13.17 (a) We are to test

$$H_0 : p_3 = p_4 \text{ vs. } H_1 : p_3 > p_4$$

(Note: Because H_1 is one-sided, the χ^2 test is not appropriate.)

We employ the test statistics

$$Z = \frac{\hat{p}_3 - \hat{p}_4}{\sqrt{\hat{p}\hat{q}}\sqrt{\frac{1}{n_1} + \frac{1}{n_2}}}$$

with large values leading to rejection of H_0. We calculate

$$\hat{p}_3 = \frac{41}{48} = .854, \hat{p}_4 = \frac{29}{50} = .580, \hat{p} = \frac{41 + 29}{48 + 50} = .714$$

$$z = \frac{.854 - .580}{\sqrt{.714 \times .286}\sqrt{\frac{1}{48} + \frac{1}{50}}} = 3.00$$

The significance probability of this observed z is

$$P - \text{value} = P[Z \geq 3.00] = .0013$$

This signifies very strong evidence in support of H_1.

(b) $\hat{p}_3 - \hat{p}_4 = .854 - .580 = .274$

Estimated $S.E. = \sqrt{\frac{.854 \times .146}{48} + \frac{.580 \times .420}{50}} = .0864$

A 95% confidence interval for $p_3 - p_4$ is

$$.274 \pm 1.96 \times .0864 = .274 \pm .169 \text{ or } (.11, .44)$$

13.19 (a) The calculations are identical.

(b) Chi-Square Test

```
Expected counts are printed below observed counts

            Open Dont Ope    Total
    1        414       386      800
             445.18    354.82

    2        532       368      900
             500.82    399.18

Total        946       754     1700

  Chi-Sq =  2.183 +  2.739 +
            1.941 +  2.435 = 9.298
DF = 1, P-Value = 0.002
```

With $\alpha = .05$ and d.f. $= 1$, we find $\chi^2_{.05} = 3.84$ so the null hypothesis of equal proportions is rejected at $\alpha = .05$. The proportions of males and females who open all of their mail are significantly different. The $P-$ value .002 greatly strengthens this conclusion.

(c) Expected counts are printed below observed counts

	C1	C2	C3	Total
1	38	15	7	60
	22.50	23.10	14.40	
2	22	32	16	70
	26.25	26.95	16.80	
3	15	30	25	70
	26.25	26.95	16.80	
Total	75	77	48	200

$$\text{ChiSq} = 10.678 + 2.840 + 3.803 +$$
$$0.688 + 0.946 + 0.038 +$$
$$4.821 + 0.345 + 4.002 = 28.162$$

df = 4

With $\alpha = .01$ and d.f. $= 4$, we find $\chi^2_{.01} = 13.28$ so the null hypothesis of equal proportions is rejected. Comparing observed and expected frequencies, we see that the bone loss is higher in the control group than the activity group.

13.21 We test the null hypothesis that the pattern of appeals decision and the type of representation are independent. The χ^2 statistic is calculated as follows:

Expected counts are printed below observed counts

	C1	C2	C3	Total
1	59	108	17	184
	74.18	98.32	11.50	
2	70	63	3	136
	54.83	72.68	8.50	
Total	129	171	20	320

```
ChiSq =  3.105 +  0.952 +  2.630 +
         4.200 +  1.288 +  3.559 = 15.734
df = 2
```

With $\alpha = .05$ and d.f. $= 2$, we find $\chi^2_{.05} = 5.99$ so the rejection region is $R : \chi^2 \geq 5.99$. Since the observed $\chi^2 = 15.734$ is in R, the null hypothesis of independence is rejected at $\alpha = .05$. We conclude that the patterns of appeals decision are significantly different between the two types of representation.

13.23 The χ^2 statistic for testing independence is calculated as follows:

```
Expected counts are printed below observed counts

              C1        C2      Total
      1       307       48        355
              308.01    46.99

      2       47         6         53
              45.99     7.01

  Total       354       54        408

     ChiSq = 0.003 + 0.022 +
             0.022 + 0.147 = 0.194
   df = 1
```

Since the observed χ^2 is less than $\chi^2_{.05} = 3.84$, the null hypothesis of independence is not rejected at $\alpha = .05$.

13.25 We test the null hypothesis of independence between union membership and attitude toward spending on social welfare. The χ^2 statistic is calculated as follows:

Expected counts are printed below observed counts

	C1	C2	C3	Total
1	112	36	28	176
	86.24	45.76	44.00	
2	84	68	72	224

```
                109.76      58.24      56.00

    Total         196        104        100        400

ChiSq =  7.695 +   2.082 +   5.818 +
         6.046 +   1.636 +   4.571 = 27.847

df = 2
```

Since the observed $\chi^2 = 27.847$ is larger than $\chi^2_{.01} = 9.21$, the null hypothesis of independence is rejected at $\alpha = .01$. We conclude that attitudes and union membership are dependent. There are significant differences between the attitudes of the union and non-union groups.

13.27 The null hypothesis that group and stopping response are independent. The cell probabilities are the product of the marginal probabilities. The χ^2 statistic is calculated as follows:

```
Expected counts are printed below observed counts

           C1        C2     Total
    1       9         9        18
           6.55     11.45

    2       8        12        20
           7.27     12.73
```

```
        3           3        14        17
                  6.18     10.82

  Total           20        35        55

    ChiSq =   0.920 +   0.526 +
              0.073 +   0.042 +
              1.638 +   0.936 = 4.134
df = 2
```

The observed χ^2 being less than $\chi^2_{.05} = 5.99$, the null hypothesis is not rejected with $\alpha = .05$. Under independence, the groups are not significantly different in their response.

13.29 Denoting the probabilities of the integers $0, 1, \ldots, 9$ by $p_0, p_1, \ldots, p_9$, the null hypothesis is:

$$H_0: \ p_0 = p_1 = \ldots = p_9 = \frac{1}{10} = .1.$$

Multiplying .1 by $n = 500$, the expected frequency of each cell is 50.

Integer	0	1	2	3	4	5	6	7	8	9	Total
O	41	58	51	61	39	56	45	35	62	52	500
E	50	50	50	50	50	50	50	50	50	50	500
$\dfrac{(O-E)^2}{E}$	1.62	1.28	.02	2.42	2.42	.72	.50	4.50	2.88	.08	16.44

$$\chi^2 = 16.44, \ \text{d.f.} \ = 9.$$

Since the observed χ^2 is smaller than $\chi^2_{.05} = 16.92$, H_0 is not rejected at $\alpha = .05$. The data do not demonstrate any bias.

13.31 Denoting the probabilities of birth in the four consecutive quarters by $p_1, p_2, p_3,$
and p_4, the null hypothesis is

$$H_0 : p_1 = \frac{2}{5}, p_2 = \frac{1}{5}, p_3 = \frac{1}{5}, p_4 = \frac{1}{5}$$

Multiplying these by $n = 300$, the expected cell frequencies are: 120, 60, 60, 60,
respectively. The calculations for the χ^2 test are shown in the following table:

	Jan-Mar	Apr-Jun	Jul-Sep	Oct-Dec	Total
O	55	29	26	41	151
E	60.4	30.2	30.2	30.2	151
$\frac{(O-E)^2}{E}$	.483	.048	.584	3.862	$4.977 = \chi^2$
					d.f.= 3

With $\alpha = .10$ and d.f. $= 3$, we find $\chi^2_{.10} = 6.25$ so the rejection region is
$R : \chi^2 \geq 6.25$. Since the observed $\chi^2 = 4.977$ is not in R, H_0 is not rejected at
$\alpha = .10$. The stated conjecture is not contradicted.

13.33 Let us denote by p_1 and p_2 the population proportions of persons having hepatitis in the two groups 'vaccinate' and 'not vaccinated', respectively. We are to test the null hypothesis $H_0 : p_1 = p_2$. The calculations for χ^2 are as follows:

```
Expected counts are printed below observed counts

            C1       C2     Total
    1       11       538       549
           41.06    507.94

    2       70       464       534
           39.94    494.06

Total       81      1002      1083

   ChiSq = 22.008 +  1.779 +
```

$$22.626 + 1.829 = 48.242$$

df = 1

We take $\alpha = .05$. For d.f. $= 1$, we find $\chi^2_{.05} = 3.84$ so the rejection region is $R: \chi^2 \geq 3.84$. Since the observed $\chi^2 = 48.242$ is in R, we reject H_0 at $\alpha = .05$. We would reject H_0 also at $\alpha = .01$ since $\chi^2_{.01} = 6.63$.

13.35 (a) The two response categories are 'free of pain' and 'not free of pain'. The frequencies of the latter category are obtained by subtracting those of the first from the corresponding 'number of patients assigned'. The 4×2 contingency table is presented here along with the calculations for the χ^2 statistic.

Expected counts are printed below observed counts

	Free	Not free	Total
1	23	30	53
	27.45	25.55	
2	30	17	47

	24.34	22.66	
3	19	32	51
	26.42	24.58	
4	29	15	44
	22.79	21.21	
Total	101	94	195

```
Chi-Sq =  0.722 +  0.776 +
          1.314 +  1.412 +
          2.082 +  2.237 +
          1.692 +  1.818 = 12.053
DF = 3, P-Value = 0.007
```

With $\alpha = .05$ and d.f. $= 3$ we find $\chi^2_{.05} = 7.81$ so the rejection region is $R: \chi^2 \geq 7.81$. Since the observed $\chi^2 = 12.053$ is in R, the null hypothesis is rejected at $\alpha = .05$. We conclude that there are significant differences in the effectiveness of the four drugs.

(b) A 90% confidence interval for a population proportion is given by

$$\hat{p} \pm 1.645\sqrt{\hat{p}\hat{q}/n}$$

We calculate

Drug 1: $\hat{p}_1 = \frac{23}{53} = .434$

$.442 \pm 1.645\sqrt{\dfrac{.434 \times .566}{53}} = .434 \pm .112$ or $(.32, .55)$

Drug 2: $\hat{p}_2 = \frac{30}{47} = .638$

$.625 \pm 1.645\sqrt{\dfrac{.638 \times .362}{47}} = .638 \pm .115$ or $(.52, .75)$

Drug 3: $\hat{p}_3 = \frac{19}{51} = .373$

$.373 \pm 1.645\sqrt{\dfrac{.373 \times .627}{51}} = .373 \pm .111$ or $(.26, .48)$

Drug 4: $\hat{p}_4 = \frac{29}{44} = .659$

$$.659 \pm 1.645 \sqrt{\frac{.659 \times .341}{44}} = .659 \pm .118 \text{ or } (.54, .78)$$

13.37 (a) In order to establish that drug 4 is more effective than drug 3, we formulate

$$H_0 : \ p_4 = p_3 \ \text{vs.} \ H_1 : \ p_4 > p_3$$

(Note: Because H_1 is one-sided, the χ^2 test is not appropriate).   We employ the test statistic

$$Z = \frac{\hat{p}_1 - \hat{p}_2}{\sqrt{\hat{p}\hat{q}}\sqrt{\frac{1}{n_1} + \frac{1}{n_2}}}$$

with large values leading to the rejection of H_0. We calculate

$$\hat{p}_4 = \frac{29}{44} = .659, \hat{p}_3 = \frac{19}{51} = .373, \hat{p} = \frac{29 + 19}{44 + 51} = .505$$

$$z = \frac{.659 - .373}{\sqrt{.505 \times .495}\sqrt{\frac{1}{44} + \frac{1}{51}}} = 2.78$$

P–value $= P[Z \geq 2.78] = .0027$.

This extremely small P–value signifies strong evidence in support of H_1.

(b) $\hat{p}_4 - \hat{p}_3 = .659 - .373 = .286$

Estimated S.E. $= \sqrt{\frac{.659 \times .341}{44} + \frac{.373 \times .627}{51}} = .0985$

A 95% confidence interval for $p_4 - p_3$ is

$$.286 \pm 1.96 \times .0985 = .286 \pm .193 \ \text{or} \ (.09, .48).$$

13.39 We test the null hypothesis that the duration of marriage is independent of the period of acquaintance before marriage. The χ^2 statistic is calculated as follows:

```
Expected counts are printed below observed counts
          C1       C2     Total
  1       11        8       19
        10.27     8.73
```

2	28	24	52
	28.11	23.89	
3	21	19	40
	21.62	18.38	
Total	60	51	111

```
ChiSq =  0.052 +  0.061 +
         0.000 +  0.000 +
         0.018 +  0.021 = 0.153
df = 2
```

With $\alpha = .05$, the tabulated value is $\chi^2_{.05} = 5.99$. The observed χ^2 is not significant. The null hypothesis of independence between period of acquaintanceship and duration of marriage is not contradicted.

13.41 We test the null hypothesis of independence of the handedness of parents and their biological offspring. The χ^2 statistic is calculated as follows:

Expected counts are printed below observed counts

	Right	Left	Total
RXR	303	37	340
	295.80	44.20	
RXL	29	9	38
	33.06	4.94	
LXR	16	6	22
	19.14	2.86	
Total	348	52	400

```
ChiSq =  0.175 +  1.173 +
         0.499 +  3.337 +
         0.515 +  3.447 = 9.146
df = 2
```

With $\alpha = .05$ and d.f. $= 2$ we find $\chi^2_{.05} = 5.99$ so the rejection region is $R : \chi^2 \geq 5.99$. Since the observed $\chi^2 = 9.146$ is in R, the null hypothesis of independence is rejected at $\alpha = .05$. The data demonstrate an association between the handedness of parents and their biological offspring. (Note: the validity of the χ^2 test is questionable here because there are two cells with expected counts less than 5.

13.43 To test the null hypothesis of independence, the χ^2 statistic is calculated as follows:

```
Expected counts are printed below observed counts

          One   Neither   Total
Innoc     27      20        47
          22.43   24.57

Unbal     36      49        85
          40.57   44.43

Total     63      69        132
```

```
     Chi-Sq =   0.930 +   0.849 +
                0.514 +   0.470 = 2.764
DF = 1, P-Value = 0.096
```

With $\alpha = .05$ and d.f. $= 1$, we find $\chi^2_{.05} = 3.84$ so the rejection region is $R : \chi^2 \geq 3.84$. Because the observed $\chi^2 = 2.764$ is not in R, the null hypothesis is rejected at $\alpha = .05$. We cannot reject the independence between type of aberration and parental carrier.

Chapter 14

ANALYSIS OF VARIANCE(ANOVA)

14.1 (a) We first find $k = 4$, $\bar{y} = 6$, $\bar{y}_1 = 1$, $\bar{y}_2 = 0$, $\bar{y}_3 = -3$, and $\bar{y}_4 = 2$. Thus,

$$
\begin{array}{cccc}
\text{Obs.} & \text{Grand mean} & \text{Tr. effect} & \text{Residuals} \\
y_{ij} & \bar{y} & \bar{y}_i - \bar{y} & y_{ij} - \bar{y}_i \\
\begin{bmatrix} 5 & 9 \\ 8 & 4 \\ 4 & 2 \\ 7 & 9 \end{bmatrix} = \begin{bmatrix} 6 & 6 \\ 6 & 6 \\ 6 & 6 \\ 6 & 6 \end{bmatrix} + \begin{bmatrix} 1 & 1 \\ 0 & 0 \\ -3 & -3 \\ 2 & 2 \end{bmatrix} + \begin{bmatrix} -2 & 2 \\ 2 & -2 \\ 1 & -1 \\ -1 & 1 \end{bmatrix}
\end{array}
$$

(b) Treatment SS $= 2(1)^2 + 2(0)^2 + 2(-3)^2 + 2(2)^2 = 28$

Residual SS $= (-2)^2 + (2)^2 + \cdots + (-1)^2 + 1^2 = 20$

Mean SS $= 8(6)^2 = 288$

Total SS $= 5^2 + 9^2 + \cdots + 9^2 = 336$

Total SS(corrected) $= (5 - 6)^2 + \cdots + (9 - 6)^2 = 336 - 288 = 48$

(c) Error d.f. $= \sum n_i - k = 2 + 2 + 2 + 2 - 4 = 4$

Treatment d.f. $= k - 1 = 4 - 1 = 3$

525

(d) The analysis-of-variance table is

ANOVA Table

Source	Sum of Squares	d.f.
Treatment	28	3
Error	20	4
Total	48	7

14.3 (a) We first find $k = 3$, $\bar{y}_1 = 3$, $\bar{y}_2 = 2$, $\bar{y}_3 = 1$, and $\bar{y} = 22/11 = 2$. Thus,

$$
\begin{array}{ccc}
\text{Obs.} & \text{Grand mean} & \text{Tr. effect} \\
y_{ij} & \bar{y} & \bar{y}_i - \bar{y}
\end{array}
$$

$$
\begin{bmatrix} 7 & 5 & 4 & 4 \\ 6 & 1 & 2 & \\ 2 & 1 & 0 & 1 \end{bmatrix}
=
\begin{bmatrix} 3 & 3 & 3 & 3 \\ 3 & 3 & 3 & \\ 3 & 3 & 3 & 3 \end{bmatrix}
+
\begin{bmatrix} 2 & 2 & 2 & 2 \\ 0 & 0 & 0 & \\ -2 & -2 & -2 & -2 \end{bmatrix}
$$

$$
\begin{array}{c}
\text{Residuals} \\
y_{ij} - \bar{y}_i
\end{array}
$$

$$
+
\begin{bmatrix} 2 & 0 & -1 & -1 \\ 3 & -2 & -1 & \\ 1 & 0 & -1 & 0 \end{bmatrix}
$$

(b) Treatment SS $= 4(2^2) + 3(0^2) + 4(-2)^2 = 32$

Residual SS $\; = 2^2 + 0^2 + \cdots + 0^2 = 22$

Mean SS $\quad = 11(3)^2 = 99$

Total SS $\quad = 7^2 + 5^2 + \cdots 0^1 + 1^2 = 153$

Total SS (corrected) $\quad = (7-3)^2 + (5-3)^2 + (4-3)^2 + (4-3)^2 + \cdots +$
$(1-3)^2 = 153 - 99 = 54$

(c) Treatment d.f. $= k - 1 = 3 - 1 = 2$.

Residual d.f. $\; = \sum n_i - k = 4 + 3 + 4 - 3 = 8$

Total d.f. $\quad = 4 + 3 + 4 - 1 = 10$

(d) The analysis-of-variance table is

ANOVA Table

Source	Sum of Squares	d.f.
Treatment	32	2
Error	22	8
Total	54	10

14.5 We first find $\bar{y}_1 = 2$, $\bar{y}_2 = 3$, $\bar{y}_3 = 6$, $\bar{y}_4 = 4$, and $\bar{y} = \frac{48}{12} = 4$. Thus,

$$
\begin{array}{ccc}
\text{Obs.} & \text{Grand mean} & \text{Tr. effect} \\
y_{ij} & \bar{y} & \bar{y}_i - \bar{y}
\end{array}
$$

$$
\begin{bmatrix}
2 & 1 & 3 & \\
1 & 5 & & \\
9 & 5 & 6 & 4 \\
3 & 4 & 5 &
\end{bmatrix}
=
\begin{bmatrix}
4 & 4 & 4 & \\
4 & 4 & & \\
4 & 4 & 4 & 4 \\
4 & 4 & 4 &
\end{bmatrix}
+
\begin{bmatrix}
-2 & -2 & -2 & \\
-1 & -1 & & \\
2 & 2 & 2 & 2 \\
0 & 0 & 0 &
\end{bmatrix}
$$

$$
\begin{array}{c}
\text{Residuals} \\
y_{ij} - \bar{y}_i
\end{array}
$$

$$
+
\begin{bmatrix}
0 & -1 & 1 & \\
-2 & 2 & & \\
3 & -1 & 0 & -2 \\
-1 & 0 & 1 &
\end{bmatrix}
$$

Treatment SS $= 3(-2)^2 + 2(-1)^2 + 4(2^2) + 3(0)^2 = 30$

Residual SS $= 0^2 + (-1)^2 + 1^2 + \cdots + 0^2 + 1^2 = 26$

Total SS $= (2-4)^2 + (1-4)^2 + (3-4)^2 + (1-4)^2 + \cdots + (5-4)^2 = 56$

Treatment d.f. $= k - 1 = 4 - 1 = 3.$

Residual d.f. $= \sum n_i - k = 3 + 2 + 4 + 3 - 4 = 8$

Total d.f. $= 3 + 2 + 4 + 3 - 1 = 11$

The analysis-of-variance table is

ANOVA Table

Source	Sum of Squares	d.f.
Treatment	30	3
Error	26	8
Total	56	11

14.7 The overall mean is

$$\bar{y} = \frac{32 \times 81.06 + 16 \times 78.56 + 16 \times 87.81}{32 + 16 + 16} = \frac{5255.84}{64} = 82.123$$

$$SS_{categories} = 32(81.06 - 82.123)^2 + 16(78.56 - 82.123)^2$$
$$+ 16(87.81 - 82.123)^2 = 756.75$$

$$SSE = 31(17.05)^2 + 15(15.43)^2 + 15(14.36)^2 = 15676.20$$

Total SS $= 756.75 + 15676.20 = 16432.95.$

ANOVA Table

Source	Sum of Squares	d.f.
Categories	756.75	2
Error	15676.20	61
Total	16432.95	63

14.9 (a) $F_{.10}(3, 5) = 3.62$

    (b) $F_{.10}(3, 10) = 2.73$

    (c) $F_{.10}(3, 15) = 2.49$

    (d) $F_{.10}(3, 30) = 2.28$

(e) Increasing the denominator d.f. decreases the upper 10th percentile.

14.11 For $\nu_1 = 5$ and $\nu_2 = 30$ in the F-table, $F_{.05}(5, 30) = 2.53$. We observe

$$F = \frac{\text{Treatment } SS/(k-1)}{SSE/(n-k)} = \frac{23/5}{56/30} = 2.46$$

so we fail to reject $H_0 : \mu_1 = \mu_2 = \mu_3 = \mu_4 = \mu_5 = \mu_6$, at level $\alpha = .05$.

14.13 We are to test the null hypothesis $H_0 : \mu_1 = \mu_2 = \mu_3$ versus the alternative hypothesis $H_1 :$ the means are not all equal. Given $\alpha = .05$, the rejection region is determined by the value $F_{.05}(2, 9) = 4.26$ obtained from the F-table. From Exercise 2.2, the observed value of F is

$$F = \frac{\text{Treatment SS}/(k-1)}{SSE/(n-k)} = \frac{312/2}{170/9} = 8.26$$

Consequently, we reject the null hypothesis that the means are equal, at the $\alpha = .05$ level of significance.

14.15 We are to test the null hypothesis $H_0 : \mu_1 = \mu_2 = \mu_3$ versus the alternative hypothesis H_1 : the means are not all equal. Given $\alpha = .05$, the rejection region is determined by the value $F_{.05}(2, 8) = 4.46$ obtained from the F-table. From

Exercise 14.3, the observed value of F is

$$F = \frac{\text{Treatment SS}/(k-1)}{\text{SSE}/(n-k)} = \frac{32/2}{22/8} = 5.82$$

Consequently, we reject the null hypothesis that the means are equal, at the

14.17 For multiple-t confidence intervals we use $t_{\alpha/2m}$ with $n-k$ d.f.

(a) $\frac{\alpha}{2m} = \frac{.05}{2(3)} = .00833$ and, with 26 d.f., $t_{.0083} = 2.559$

(b) $\frac{\alpha}{2m} = \frac{.05}{2(5)} = .005$ and, with df $= 26$, $t_{.005} = 2.779$.

14.19 The error d.f. $= (20 + 18 + 24 + 18 - 4) = 66$. The t intervals use $t_{.025} = 2.00$ and the multiple-t intervals use $\frac{\alpha}{2m} = \frac{.05}{2(6)} = .00417$ so we extrapolate $t_{.00417} = 2.7$.

	t-interval	multiple-t interval
$\mu_1 - \mu_2$:	2.1 ± 2.08	2.1 ± 2.81
$\mu_1 - \mu_3$:	$.5 \pm 1.94$	$.5 \pm 2.62$
$\mu_1 - \mu_4$:	4.0 ± 2.68	4.0 ± 3.61
$\mu_2 - \mu_3$:	-1.6 ± 1.99	-1.6 ± 2.69
$\mu_2 - \mu_4$:	1.9 ± 2.71	1.9 ± 3.67
$\mu_3 - \mu_4$:	3.5 ± 2.60	3.5 ± 3.53

Only μ_1 and μ_4 differ according to the multiple t-intervals.

14.21 The t interval is $\bar{Y}_i - \bar{Y}_{i'} \pm t_{\alpha/2} S \sqrt{\frac{1}{n_i} + \frac{1}{n_{i'}}}$

The multiple-t interval is $\bar{Y}_i - \bar{Y}_{i'} \pm t_{\alpha/2m} S \sqrt{\frac{1}{n_i} + \frac{1}{n_{i'}}}$ so the ratio of lengths

$$\frac{2t_{\alpha/2} S \sqrt{\frac{1}{n_i} + \frac{1}{n_{i'}}}}{2t_{\alpha/2m} S \sqrt{\frac{1}{n_i} + \frac{1}{n_{i'}}}} = \frac{t_{\alpha/2}}{t_{\alpha/2m}}$$

does not depend on the data. For $m = 10$ and $\alpha = .10$, the ratio is

$$\frac{t_{.05}}{t_{.005}} = \frac{1.753}{2.947} = .595 \text{ for 15 d.f.}$$

14.23 (a) We first find $k = 3$, $\bar{y}_{..} = 8$, $\bar{y}_{1.} = 7$, $\bar{y}_{2.} = 6$, $\bar{y}_{3.} = 11$. Also $b = 4$, $\bar{y}_{.1} = 11$, $\bar{y}_{.2} = 8$, $\bar{y}_{.3} = 9$ and $\bar{y}_{.4} = 4$.

Thus,

$$
\begin{array}{ccccc}
\text{Obs.} & & \text{Grand mean} & & \text{Tr. effect} \\
y_{ij} & & \bar{y}_{..} & & \bar{y}_{i.} - \bar{y}_{..} \\
\begin{bmatrix} 11 & 10 & 7 & 0 \\ 7 & 8 & 7 & 2 \\ 15 & 6 & 13 & 10 \end{bmatrix}
& = &
\begin{bmatrix} 8 & 8 & 8 & 8 \\ 8 & 8 & 8 & 8 \\ 8 & 8 & 8 & 8 \end{bmatrix}
& + &
\begin{bmatrix} -1 & -1 & -1 & -1 \\ -2 & -2 & -2 & -2 \\ 3 & 3 & 3 & 3 \end{bmatrix}
\end{array}
$$

$$
\begin{array}{ccccc}
& \text{Bl. effect} & & \text{Error} \\
& \bar{y}_{.j} - \bar{y}_{..} & & y_{ij} - \bar{y}_{i.} - \bar{y}_{.j} + \bar{y}_{..} \\
+ &
\begin{bmatrix} 3 & 0 & 1 & -4 \\ 3 & 0 & 1 & -4 \\ 3 & 0 & 1 & -4 \end{bmatrix}
& + &
\begin{bmatrix} 1 & 3 & -1 & -3 \\ -2 & 2 & 0 & 0 \\ 1 & -5 & 1 & 3 \end{bmatrix}
\end{array}
$$

(b) The sums of squares are

$$
\begin{aligned}
\text{Treatment SS} &= 4(-1)^2 + 4(-2)^2 + 4(3)^2 = 56 \\
\text{Block SS} &= 3(3)^2 + 3(0)^2 + 3(1)^2 + 3(-4)^2 = 78 \\
\text{Residual SS} &= 1^2 + 3^2 + (-1)^2 + \cdots + 1^2 + 3^2 = 64 \\
\text{Mean SS} &= 12(8)^2 = 768 \\
\text{Total SS} &= 11^2 + 10^2 + 7^2 + \cdots + 13^2 + 10^2 = 966 \\
\text{Total SS (corrected)} &= 11^2 + 10^2 + 7^2 + \cdots + 13^2 + 10^2 - 12(8)^2 \\
&= 966 - 768 = 198
\end{aligned}
$$

(c) Treatment d.f. $= k - 1 = 2$

Block d.f. $= b - 1 = 3$

Residual d.f. $= (k - 1)(b - 1) = 2 \cdot 3 = 6$

14.25 The null hypothesis is that the three treatment population means are the same. The alternative is that they are not the same. The analysis-of-variance table is

Source of variation	Degrees of freedom	Sum of squares	Mean square	F
Treatments	2	56	28	2.62
Blocks	3	78	26	2.44
Error	6	64	10.667	
Total	11	198		

Since the critical value at the 0.05 level for an F distribution with 2 and 6 degrees of freedom is 5.14, we fail to reject the null hypothesis of equal treatment means.

Since $F_{.05}$ with 3 and 6 degrees of freedom is 4.76, the block effect is not significant.

14.27 (a) At each baking, select a loaf of bread from each recipe and randomize the position of the loaves in the oven.

  (b) The Grand mean is $\bar{y}_{..} = \frac{1}{15}(.95 + .71 + \ldots + .44) = \frac{10.57}{15} = .7047$.

The Block means are .7833, .7967, .6133, .7233, and .6067.

The recipe means are .796, .708, and .610.

$$SS_B = 3[(.7833 - .7074)^2 + (.7967 - .7047)^2 + (.6133 - .7047)^2$$
$$+(.7233 - .7047)^2 + (.6067 - .7047)^2] = .0988$$

$$SS_T = 5[(.796 - .7047)^2 + (.708 - .7047)^2 + (.610 - .7047)^2] = .0866$$

Total $SS = (.95 - .7047)^2 + \ldots + (.44 - .7047)^2 = .2280$, hence

$$SSE = .2280 - (.0866 + .0988) = .0426.$$

The ANOVA table is

Source	Sum of squares	d.f.	Mean square	F-ratio
Treatments	.0866	2	.0433	8.17
Blocks	.0988	4	.0247	4.66
Residual	.0426	8	.0053	
Total	.2280	14		

Since $F_{.05}(2,8) = 4.46 < 8.17$ we conclude that a significant treatment difference is indicated by the data. Also, $F_{.05}(4,8) = 3.84 < 4.66$ so the block effects are significant.

14.29 (a) The Grand mean is

$$\bar{y}_{..} = \frac{1}{36}(19.09 + 16.28 + \ldots + 21.58) = \frac{653.96}{36} = 18.1656.$$

The block means are 17.7533, 18.78, 18.2367, 17.94, 17.6867, and 18.5967.

The variety means are 19.6683, 17.1083, 17.2683, 17.7, 16.0767, and 21.177.

$$SS_B = 6[(17.7533 - 18.1656)^2 + (18.78 - 18.1656)^2$$

$$+ \ldots + (18.5967 - 18.1656)^2] = 6.112$$

$$SS_T = 6[(19.6683 - 18.1656)^2 + (17.1083 - 18.1656)^2$$

$$+ \ldots + (21.1717 - 18.1656)^2] = 106.788$$

$$SS = (19.09 - 18.1656)^2 + (16.28 - 18.1656)^2 + \ldots + (21.58 - 18.1656)^2$$

$$= 117.898$$

hence

$$SSE = 117.898 - (6.112 + 106.788) = 4.998.$$

The ANOVA table is

Source	Sum of squares	d.f.	Mean square	F-ratio
Treatments	106.788	5	21.358	106.79
Blocks	6.112	5	1.222	6.11
Residual	4.998	25	.200	
Total	117.898	35		

Since $F_{.05}(5, 25) = 2.603 < 106.79$ we conclude that a highly significant treatment difference is indicated by the data. Also, $F_{.05}(5, 25) < 6.11$ so the block effects are significant.

(b) Array of residuals

$$\begin{bmatrix} -.166 & -.416 & -.546 & .212 & .586 & .331 \\ .007 & .157 & .287 & -.264 & .229 & -.416 \\ .571 & -.299 & .041 & -.181 & -.268 & .137 \\ .157 & .687 & .487 & .166 & -1.071 & -.426 \\ -.569 & .091 & -.449 & .159 & .372 & .397 \\ .001 & -.219 & .181 & -.091 & .151 & -.023 \end{bmatrix}$$

The dot diagrams are given in Figure 14.2. Note that -1.071 is a possible outlier. The normal scores plot is given in Figure 14.3 and the same point is a little low.

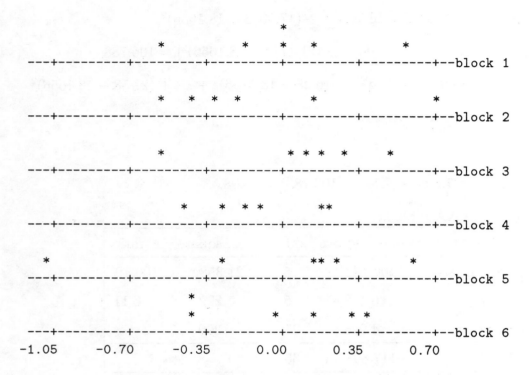

Figure 14.2: Dot diagrams for Exercise 14.29.

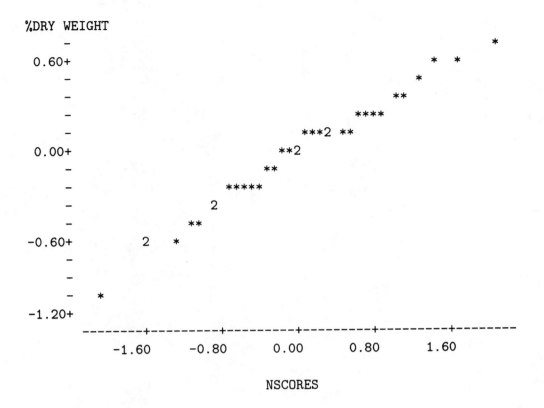

Figure 14.3: Normal-scores plot for Exercise 14.29.

14.31 $k = 3$ we first find $\bar{y}_1 = 19, \bar{y}_2 = 13, \bar{y}_3 = 14, \bar{y} = 225/15 = 15$

$$
\begin{matrix}
y_{ij} \\
\begin{bmatrix}
19 & 18 & 21 & 18 & & \\
16 & 11 & 13 & 14 & 11 & \\
13 & 16 & 18 & 11 & 15 & 11
\end{bmatrix}
\end{matrix}
=
\begin{matrix}
\bar{y} \\
\begin{bmatrix}
15 & 15 & 15 & 15 & & \\
15 & 15 & 15 & 15 & 15 & \\
15 & 15 & 15 & 15 & 15 & 15
\end{bmatrix}
\end{matrix}
$$

$$
+
\begin{matrix}
(\bar{y}_i - \bar{y}) \\
\begin{bmatrix}
4 & 4 & 4 & 4 & & \\
-2 & -2 & -2 & -2 & -2 & \\
-1 & -1 & -1 & -1 & -1 & -1
\end{bmatrix}
\end{matrix}
+
\begin{matrix}
(y_{ij} - \bar{y}_i) \\
\begin{bmatrix}
0 & -1 & 2 & -1 & & \\
3 & -2 & 0 & 1 & -2 & \\
-1 & 2 & 4 & -3 & 1 & -3
\end{bmatrix}
\end{matrix}
$$

14.33   (a) $F_{.05}(7, 13) = 2.83$ (b) $F_{.05}(7, 20) = 2.51$

(c) $F_{.10}(7, 12) = 2.28$.

14.35

One-way Analysis of Variance

Analysis of Variance
Source   DF   SS   MS   F   P
Factor   2   5.279   2.640   20.85   0.000
Error   14   1.772   0.127
Total   16   7.052
      Individual 95% CIs For Mean

```
                                    Based on Pooled StDev

Level      N    Mean     StDev   --+---------+---------+---------+-
Treat 1    5   1.4600    0.1306  (----*-----)
Treat 2    6   1.5967    0.4357   (-----*----)
Treat 3    6   2.6950    0.3886                          (----*----)
                                 --+---------+---------+---------+-
Pooled StDev =    0.3558          1.20      1.80      2.40      3.00
                                            Moisture
```

14.37 (a) We first find $k = 3$, $\bar{y}_{..} = 8$, $\bar{y}_{1.} = 6$, $\bar{y}_{2.} = 7$, $\bar{y}_{3.} = 11$. Also $b = 4$, $\bar{y}_{.1} = 7$, $\bar{y}_{.2} = 12$, $\bar{y}_{.3} = 3$ and $\bar{y}_{.4} = 10$.

Thus,

$$
\begin{array}{ccc}
\text{Obs.} & \text{Grand mean} & \text{Tr. effect} \\
y_{ij} & \bar{y}_{..} & \bar{y}_{i.} - \bar{y}_{..}
\end{array}
$$

$$
\begin{bmatrix}
8 & 9 & 1 & 6 \\
5 & 12 & 0 & 11 \\
8 & 15 & 8 & 13
\end{bmatrix}
=
\begin{bmatrix}
8 & 8 & 8 & 8 \\
8 & 8 & 8 & 8 \\
8 & 8 & 8 & 8
\end{bmatrix}
+
\begin{bmatrix}
-2 & -2 & -2 & -2 \\
-1 & -1 & -1 & -1 \\
3 & 3 & 3 & 3
\end{bmatrix}
$$

$$
\begin{array}{cc}
\text{Bl. effect} & \text{Error} \\
\bar{y}_{.j} - \bar{y}_{..} & y_{ij} - \bar{y}_{i.} - \bar{y}_{.j} + \bar{y}_{..}
\end{array}
$$

$$
+
\begin{bmatrix}
-1 & 4 & -5 & 2 \\
-1 & 4 & -5 & 2 \\
-1 & 4 & -5 & 2
\end{bmatrix}
+
\begin{bmatrix}
3 & -1 & 0 & -2 \\
-1 & 1 & -2 & 2 \\
-2 & 0 & 2 & 0
\end{bmatrix}
$$

(b) The sums of squares are

$$
\begin{aligned}
\text{Treatment SS} &= 4(-2)^2 + 4(-1)^2 + 4(3)^2 = 56 \\
\text{Block SS} &= 3(-1)^2 + 3(4)^2 + 3(-5)^2 + 3(2)^2 = 138 \\
\text{Residual SS} &= 3^2 + (-1)^2 + 0^2 + \cdots + 2^2 + 0^2 = 32 \\
\text{Total SS} &= 8^2 + 9^2 + 1^2 + \cdots + 8^2 + 13^2 - 12(8)^2 = 226
\end{aligned}
$$

(c) The $k = 3$ rows of the treatment array sum to zero and the $b = 4$ columns of the block array sum to zero. All of the entries of the Error array sum to zero as do the columns and the rows. Consequently,

$$
\text{Treatment d.f.} = k - 1 = 2
$$

$$\text{Block d.f.} = b - 1 = 3$$

$$\text{Residual d.f.} = (k - 1)(b - 1) = 2 \cdot 3 = 6$$

Chapter 15

NONPARAMETRIC INFERENCE

15.1 (a) Rank collections for treatment B with sample sizes $n_A = 4$ and $n_B = 2$

Rank of B	Rank Sum W_B	Probability
1, 2	3	1/15
1, 3	4	1/15
1, 4	5	1/15
1, 5	6	1/15
1, 6	7	1/15
2, 3	5	1/15
2, 4	6	1/15
2, 5	7	1/15
2, 6	8	1/15
3, 4	7	1/15
3, 5	8	1/15
3, 6	9	1/15
4, 5	9	1/15
4, 6	10	1/15
5, 6	11	1/15
	Total	1

When the two samples come from the same population, every pair of integers out of {1,2,3,4,5,6} is equally likely to be the ranks for the two B measurements. There are $\binom{6}{2} = 15$ potential pairs so that each collection of possible ranks has a probability of 1/15.

(b) Both rank collections $\{1, 4\}$ and $\{2, 3\}$ have $W_B = 5$ so

$$P[W_B = 5] = 1/15 + 1/15 = 2/15.$$

Continuing, we obtain the distribution of W_B.

Values of W_B	3	4	5	6	7	8	9	10	11
Probability	$\frac{1}{15}$	$\frac{1}{15}$	$\frac{2}{15}$	$\frac{2}{15}$	$\frac{3}{15}$	$\frac{2}{15}$	$\frac{2}{15}$	$\frac{1}{15}$	$\frac{1}{15}$

These values agree with the tabulated entries of Table 7.

15.3 (a) Smaller sample size = 5, larger sample size = 6, so $P[W_S \geq 39] = .063$

(b) Smaller sample size $= 4$, larger sample size $= 6$, so $P[W_S \leq 15] = .086$

(c) With smaller sample size $= 7$, larger sample size $= 7$, $P[W_S \geq 66] = .049$ so $c = 66$.

15.5 (a)

Combined sample ordered observations	2.1	2.7	3.2	3.7	5.3
Ranks	1	2	3	4	5
Treatment	A	B	B	A	A

We find

$$W_A = 1 + 4 + 5 = 10.$$

(b) Since n_B is the smaller sample size,

$$W_S = 2 + 3 = 5.$$

15.7 We test H_0 : the populations are identical versus H_1 : the populations are different. Let W_S = rank sum of phosphate for the Chester White breed. The alternative is two-sided. From Appendix Table 9 we find with smaller size = 8 = larger size, $P[W_S \geq 87] = .025 = P[W_S \leq 49]$.

The combined ordered observations, with the Chester White underlined are

Ordered observations	47	48	57	58	65	75	78	79
Ranks	1	2	3	4	5	6	7	8

Ordered observations	97	99	110	162	172	182	220	230
Ranks	9	10	11	12	13	14	15	16

We find

$$W_S = 1 + 2 + 3 + 5 + 6 + 9 + 10 + 11 = 47$$

We conclude that the serum phosphate level is significantly different for the two breeds at the level $\alpha = .05$.

15.9 We test H_0 : the populations are identical versus H_1 populations are different. Let W_S = rank sum for treatment 2. The alternative is two-sided. From

Appendix Table 9 we find with smaller size $= 7$ and larger size $= 8$,

$P[W_S \geq 73] = .027 = P[W_S \leq 39]$. The sample sizes $n_A = 8$ and $n_B = 7$.

Combined ordered values	18	25	28	29	30	31	36	37	38	40	41	43	46	49	56
Ranks	1	2	3	4	5	6	7	8	9	10	11	12	13	14	15

$$W_S = 2 + 5 + 7 + 8 + 11 + 14 + 15 = 62$$

Consequently, we fail to reject the null hypothesis at level $\alpha = .054$.

15.11 (a) The configuration that most supports the alternative hypothesis is
       $BBBBBBBBBA$ where the rank of the single A observation is 10.

  (b) There are 10 possible ranks (positions) for the single A and these are
      equally likely. Therefore $P[W_A = 10] = .1$.

  (c) The single most extreme outcome has probability .1. An α of .05 cannot
      be achieved unless, whenever $W_A = 10$, we are willing to flip a coin to
      decide whether or not H_0 should be rejected.

15.13 The alternative is one-sided. We test H_0 : the is no difference in treatments against $H_1 : P[+] > .5$. Let $S =$ number of positive signs among the differences.

The rejection region is $R : S \geq c$. There are 13 positive values out of 18. From Appendix Table 2 with $n = 18$ we find $P[S \geq 13] = .048$ Since the observed value is $S = 13$, we reject that H_0 , at level $\alpha = .048$, in favor of $P[+] > .5$ or more than half of the population prefer recipe A.

15.15 The alternative is two-sided. We test H_0 : no difference in treatments against $H_1 :$ $P[+] \neq .5$. Let $S =$ number of positive signs among the differences. The rejection region is $R :$ $S \leq c_1$ or $S \geq c_2$. Since there are 7 ties the effective sample size is $25 - 7 = 18$. From Appendix Table 2 with $n = 18$, we find $P[S \leq 4] = .015 = P[S \geq 14]$. Since the observed value is $S = 11$, we conclude at $\alpha = .030$ that the data do not show a significant difference of opinion between husbands and wives.

15.17 (a) $P[T^+ \geq 54] = .034$,   (b) $P[T^+ \leq 32] = .060$

(c) Since $P[T^+ \geq 79] = .052, c = 79$

15.19 (a) The alternative is two-sided. The rejection region is $R : \ T^+ \leq c_1$ or $T^+ \geq c_2$. From the table for the signed rank statistic, with $n = 6$ we find $P[T^+ \leq 0] = .016 = P[T^+ \geq 21]$. We calculate

Restaurant	1	2	3	4	5	6
Critic 1	6.1	5.2	8.9	7.4	4.3	9.7
Critic 2	7.3	5.5	9.1	7.0	5.1	9.8
difference $C2 - C1$	1.2	0.3	0.2	−0.4	0.8	0.1
sign	+	+	+	−	+	+
rank ordered $|C2 - C1|$	6	3	2	4	5	1

Since the observed value is $T^+ = 1 + 2 + 3 + 5 + 6 = 17$, we fail to reject H_0 at level $\alpha = .032$.

(b) We must consider extreme values in both tails. From the table with $n = 6$ we find $P[T^+ \leq 4] = .109 = P[T^+ \geq 17]$ so the observed value $T^+ = 17$

has significance probability $2(.109) = .218$.

15.21 (a)

	Ranks	T^+	Probability
	1, 2, 3		
	+, +, +	6	.125
	+, +, −	3	.125
	+, −, +	4	.125
Signs	−, +, +	5	.125
	+, −, −	1	.125
	−, +, −	2	.125
	−, −, +	3	.125
	−, −, −	0	.125

(b)

Values of T^+	0	1	2	3	4	5	6
Probability	.125	.125	.125	.250	.125	.125	.125

From Appendix Table 10, $n = 3$, we can see that the tail probabilities agree.

15.23 (a) The alternative is one-sided. We test H_0 : the is no difference in treatments against H_1 : $P[+] > .5$. There are two ties among the differences so the effective sample size is $n = 13$. Let S = number of positive signs among the 13 differences $before - after$. The rejection region is R : $S > c$. From Appendix Table 2 with $n = 13$ we find $P[S \geq 11] = .046$.

Since the observed value is $S = 11$, we reject H_0 at level $\alpha = .046$.

(b) The alternative is one-sided so we reject for large values of T^+. Using the 0 differences for ranking and average rank for the other ties, we calculate

Ordered absolute value of differences	0	0	2	4	4	6	8	8
Ranks	1.5	1.5	3	4.5	4.5	6	7.5	7.5
Signs			+	+	+	+	+	+

Ordered absolute value of differences	10	10	10	18	18	26	32
Ranks	10	10	10	12.5	12.5	14	15
Signs	+	−	+	+	+	−	+

The observed value is $T^+ = 93$. There are ties so the most accurate answer would be to find the distribution of T^+ under the tied structure(See Lehmann reference in text). To give a more expedient answer, we take a conservative approach and assign the difference -10 rank 11 and the two positive differences ranks 9 and 10. The resulting observed value of T^+ is then 92. If there were no ties, R : $T^+ \geq 990$ since $P[T^+ \geq 90] = .047$ for

$n = 15$. Consequently, we reject H_0 and conclude that the mean blood pressure has been reduced.

15.25 We calculate

x	3.1	5.4	4.7
y	2.8	3.5	4.6
Ranks R_i	1	3	2
Ranks X_i	1	2	3

$$r_{S_P} = \frac{\sum_{i=1}^{3}(R_i - \frac{3+1}{2})(S_i - \frac{3+1}{2})}{\frac{3(3^2-1)}{12}}$$

$$= \frac{(1-2)(1-2) + (3-2)(2-2) + (2-2)(3-2)}{2} = 0.5$$

15.27 We are to test H_0 : independence against a two-sided alternative.

Student	1	2	3	4	5	6	7	8	9	10
Dexterity	23	29	45	36	49	41	30	15	42	38
Aggression	45	48	16	28	38	21	36	18	31	37
Ranks R_i	2	3	9	5	10	7	4	1	8	6
Ranks S_i	9	10	1	4	8	3	6	2	5	7

$$r_{S_P} = \frac{\sum_{i=1}^{10}(R_i - \frac{10+1}{2})(S_i - \frac{10+1}{2})}{\frac{10(10^2-1)}{12}}$$

$$= \frac{(2-5.5)(9-5.5) + (3-5.5)(10-5.5) + \cdots + (6-5.5)(7-5.5)}{76.5}$$

$$= \frac{16.5}{82.5} = -.20$$

Even though $n = 10$ is not large, we approximate that $\sqrt{n-1}\,r_{Sp}$ is approximately standard normal. Since the z value $\sqrt{9}(-.20) = -.60$ is not negative enough, to reject the null hypothesis of independence.

15.29

Combined sample ordered observations	32	43	67	81	90	99
Ranks	1	2	3	4	5	6
Treatment	A	B	B	A	A	B

We find

$$W_A = 1 + 4 + 5 = 10.$$

15.31 (a) Rank collections for treatment A with sample sizes $n_A = 3$ and $n_B = 2$

Rank of A	Rank Sum W_A	Probability
1, 2, 3	6	1/10
1, 2, 4	7	1/10
1, 2, 5	8	1/10
1, 3, 4	8	1/10
1, 3, 5	9	1/10
1, 4, 5	10	1/10
2, 3, 4	9	1/10
2, 3, 5	10	1/10
2, 4, 5	11	1/10
3, 4, 5	12	1/10
	Total	1

When the two samples come from the same population, every triple $\{1,2,3,4,5\}$ is equally likely to be the ranks for the three A measurements. There are $\binom{5}{3} = 10$ potential triples so that each collection of possible ranks has a probability of $1/10$.

(b) Both rank collections $\{1, 2, 5\}$ and $\{1, 3, 4\}$ have $W_A = 8$ so

$$P[W_A = 8] = 1/10 + 1/10 = .2.$$

Continuing, we obtain the distribution of W_A.

Values of W_A	6	7	8	9	10	11	12
Probability	.1	.1	.2	.2	.2	.1	.1

15.33 (a) $P[T^+ \geq 28] = .098,$ (b) $P[T^+ \leq 5] = .020$

(c) Since $P[T^+ \leq 21] = .047$, we have $c = 21$.

15.35 We test H_0 : populations A and B are identical versus H_1 : they are different.

Combined ordered values	95	98	100	103	104	105	116	127	131	137
Ranks	1	2	3	4	5	6	7	8	9	10

Combined ordered values	140	149	150	151	155	164	167	178	179
Ranks	11	12	13	14	15	16	17	18	19

The rank sum of method 2 (the smaller sample) is

$$W_S = 1 + 2 + 3 + 4 + 6 + 7 + 8 + 9 + 11 = 51.$$

Referring to Appendix Table 9 with smaller sample size = 9 and larger sample size = 10 we find $P[W_S \leq 69] = .047 = P[W_S \geq 111]$. Since the observed value = 51 < 69, we conclude there is a significant difference at level $\alpha = .047 + .047 = .094$. In fact, the null hypothesis would be rejected even for α much smaller than .018.

15.37 (a) For the Chester White,

Calcium	116	112	82	63	117	69	79	87
Phosphate	47	48	57	75	65	99	97	110
Ranks R_i	7	6	4	1	8	2	3	5
Ranks S_i	1	2	3	5	4	7	6	8

$$r_{S_P} = \frac{\sum_{i=1}^{3}(R_i - \frac{8+1}{2})(S_i - \frac{8+1}{2})}{\frac{8(8^2 - 1)}{12}} = \frac{-22}{42} = -0.524$$

(b) We test the null hypothesis H_0 : independence against a two-sided alternative. The value of the test statistic is $\sqrt{n-1}(-.524) = -1.386$. Consequently, we fail to reject the hypothesis of independence.

(c) If we approximate that $\sqrt{n-1}\,r_{S_p}$ is nearly standard normal, the level of significance is

$$\alpha = P[Z < -1.96] + P[Z > 1.96] = .025 + .025 = .05$$

However, $n = 7$ may not be large enough for a good approximation.

15.39 (a) We calculate

x	10	7	8
y	15	13	9
difference $d = y - x$	5	6	1
sign	+	+	+
Ranks absolute d	2	3	1

$$S = 1 + 1 + 1 = 3$$

(b)

$$T^+ = 1 + 2 + 3 = 6$$

Appendix

Formula of Total Probability, Bayes' Rule, and Applications

Formula of Total Probability, Bayes' Rule, and Applications

Recall that for any event A, the pair of events A and $\overline{A}$ has an intersection that is empty, whereas the union $A \cup \overline{A}$ represents the total population of interest. In fact, this pair of events $\{A, \overline{A}\}$ is a special case of a *partition* of the sample space, hereinafter denoted by S.

1. *Partition of Sample Space and Formula of Total Probability.*

Definition of Partition. A collection of events $\{S_1, S_2, \cdots, S_n\}$ of a certain sample space (or population) S is called a **partition** if
    (i) $S_1, S_2, \cdots, S_n$ are *mutually exclusive events*;
    (ii) $S_1 \cup S_2 \cup \cdots \cup S_n = S$.

Illustrative Example. In the nineteenth century G. Mendel conducted a famous experiment that led to the first announcement of elementary genetic principles. He bred hybrid strains of peas and simultaneously observed the color (green or yellow) and smoothness (round or wrinkled) of the offspring peas. If S denotes the set of all peas involved in the pea-breeding experiment, and
    S_1 denotes the subpopulation of round and green peas;
    S_2 denotes the subpopulation of round and yellow peas;
    S_3 denotes the subpopulation of wrinkled and green peas;
    S_4 denotes the subpopulation of wrinkled and yellow peas;
then $\{S_1, S_2, S_3, S_4\}$ represents a partition of S.

Formula of Total Probability.

Assume that the set of events $\{S_1, S_2, \cdots, S_n\}$ constitutes a partition of the sample space S. Assume that for every i, $1 \leq i \leq n$,
$$P(S_i) > 0.$$
Then for any event A, we have
(1)
$$P(A) = \sum_{i=1}^{n} P(S_i) \cdot P(A \mid S_i).$$

Proof. It follows from the multiplication law of probability that for every i, $1 \leq i \leq n$,

$$P(S_i) \cdot P(A \mid S_i) = P(AS_i).$$

On the other hand, since the events S_1, S_2, $\cdots$, S_n are mutually exclusive, we have that the events AS_1, AS_2, $\cdots$ AS_n are also mutually exclusive. In addition note that

(2) $$AS_1 \cup AS_2 \cup \cdots \cup AS_n = A.$$

It would be instructive if the student tries to verify (2) by use of a Venn diagram. Then an application of the addition law of probability to (2) gives (1).

Illustrative Example. A diagnostic test for a certain disease is known to be 95% accurate, i.e., if a person has the disease, the test will detect it with probability 0.95. Also, if the person does not have the disease, the test will report that they do not have it with the same probability 0.95. In addition, it is known from previous data that only 1% of the population has this particular disease. What is the probability that a particular person chosen at random will be tested positive?

Solution. Let

T^+ denote the event that a person is tested positive;

T^- denote the event that a person is tested negatively;

D denote the event that a person has the disease.

Then it follows from the above stated conditions that

$$P(D) = 0.01, \quad P(\overline{D}) = 0.99,$$

$$P(T^+ \mid D) = 0.95, \text{ and } P(T^- \mid \overline{D}) = 0.95.$$

In particular, since

$$\underbrace{P(T^- \mid \overline{D})}_{0.95} + P(T^+ \mid \overline{D}) = 1,$$

we have that

$$P(T^+ \mid \overline{D}) = 0.05.$$

Now we apply the formula of total probability from (1) with $A = T^+$, $n = 2$, $S_1 = D$, and $S_2 = \overline{D}$, to obtain

$$P\left(T^+\right) = P\left(T^+ \mid D\right) \cdot P(D) + P\left(T^+ \mid \overline{D}\right) \cdot P\left(\overline{D}\right)$$

$$= (0.95) \cdot (0.01) + (0.05) \cdot (0.99)$$

$$\approx 0.059.$$

2. Bayes' Rule.

This important rule enables one to compute a conditional probability when the original condition now becomes the event of interest.

Assume that the set $\left\{ S_1, S_2, \cdots, S_n \right\}$ constitutes a partition of the sample space S. Assume that for each i, $1 \le i \le n$,

$$P(S_i) > 0.$$

Fix any event A. Then for any given j, $1 \le j \le n$,

(3)
$$P\left(S_j \mid A\right) = \frac{P\left(S_j\right) \cdot P\left(A \mid S_j\right)}{\sum\limits_{i=1}^{n} P\left(S_i\right) \cdot P\left(A \mid S_i\right)}.$$

The key thing to note about Bayes' theorem is that the information that will be given in a problem will be the conditional probabilities $P(A \mid S_i)$, $1 \le i \le n$, that appear on the right-hand side of the equation, whereas what is sought is one of the conditional probabilities, $P\left(S_j \mid A\right)$, where the events S_j and A are "reversed" from the given information. I.e., given that A occurred, what is the probability it happened "through S_j".

Proof of Bayes' theorem. Note that by the multiplication law of probability, the numerator of the fraction on the right-hand side of (3) can be rewritten as

$$P\left(S_j\right) \cdot P\left(A \mid S_j\right) = P\left(AS_j\right).$$

At the same time, by the formula of total probability (formula (1) above), the denominator of the fraction on the right-hand side of (3) is equal to

$$\sum_{i=1}^{n} P\left(S_i\right) \cdot P\left(A \mid S_i\right) = P(A).$$

Hence, the right-hand side of (3) is equal to

$$\frac{P\left(AS_j\right)}{P(A)} = P\left(S_j \mid A\right)$$

by the definition of conditional probability.

3. *Examples.*

Illustrative Example 1. It is quite common that different illnesses produce similar or even identical symptoms. Suppose that any one of the illnesses X, Y, or Z lead to the same set of symptoms, hereafter denoted as U. For simplicity assume that the illnesses X, Y, and Z are mutually exclusive and that there are no other illnesses leading to the same set of symptoms. Suppose the probabilities of contracting these three illnesses are:

$$P(X) = 0.03, \qquad P(Y) = 0.01, \qquad P(Z) = 0.02,$$

and that the chances of developing the set of symptoms U, given a specific illness are:

$$P(U \mid X) = 0.85, \qquad P(U \mid Y) = 0.92 \qquad P(U \mid Z) = 0.80.$$

If a sick person develops the set of symptoms U, what are the chances he or she has illness X ?

Solution. First note that the set of events X, Y, and Z together do not represent a partition. Therefore, define H to be the event of not suffering from any of X, Y, or Z, i.e., the complement of the union of X, Y, and Z,

$$H = \overline{X \cup Y \cup Z}.$$

Then we have

$$P(H) = 1 - P(X) - P(Y) - P(Z)$$
$$= 1 - 0.03 - 0.01 - 0.02$$
$$= 0.94.$$

However,

$$P(U \mid H) = 0.$$

Applying Bayes' rule yields that the conditional probability that given the symptoms U, that a person indeed has the illness X, viz., $P(X \mid U)$, is

$$P(X \mid U) = \frac{P(U \mid X) \cdot P(X)}{P(U \mid X) \cdot P(X) + P(U \mid Y) \cdot P(Y) + P(U \mid Z) \cdot P(Z) + P(U \mid H) \cdot P(H)}$$

$$= \frac{(0.85) \cdot (0.03)}{(0.85) \cdot (0.03) + (0.92) \cdot (0.01) + (0.80) \cdot (0.02) + 0}$$

$$= 0.5029.$$

Note that the data given at the outset of the problem above involved the conditional probabilities of "U given X, U given Y, U given Z, and U given H", but what was sought was the conditional probability of "X given U", which involved the reverse of the conditions of the given data in the problem. This is the prototypical situation for the application of Bayes' theorem.

Illustrative Example 2. In this example we consider a situation somewhat like the earlier example above on pp. 2 and 3 of this insert following the formula of total probability. Suppose we are concerned with medically testing for leukemia. Let

T^+ denote the event that the test is positive, suggesting the person has leukemia;

T^- denote the event that the test is negative, suggesting the person does not have leukemia;

L denote the event that the person tested has leukemia;

$\bar{L}$ denote the event that the person tested does not have leukemia.

It is the case that the medical test for leukemia is not perfectly accurate. Most of the time, if one has leukemia, the test will be positive. Past records indicate that $P(T^+ \mid L) = 0.98$. Similarly, if one does not have leukemia, the test is usually negative. Again, it is known that $P(T^- \mid \bar{L}) = 0.99$. All this notwithstanding, there are people who sometimes test positively, but do not , in fact, have the disease; and some who test negatively, but do indeed have the disease. If we also know that $P(L) = 0.000001$, find:

(a) the probability $P(\bar{L} \mid T^+)$ of a false positive test;

and

(b) the probability $P(L \mid T^-)$ of a false negative test.

Solution. (a) Note again that in this problem we are given the conditional probabilities of "T^+ given L and T^- given $\bar{L}$", but are asked to find the conditional probabilities that have the T^+, T^- events and the L, $\bar{L}$ events "reversed". Hence we employ Bayes' rule. This yields

$$P(\bar{L} \mid T^+) = \frac{P(T^+ \mid \bar{L}) \cdot P(\bar{L})}{P(T^+ \mid \bar{L}) \cdot P(\bar{L}) + P(T^+ \mid L) \cdot P(L)} .$$

Since

$$P(\bar{L}) = 1 - P(L) = 0.999999 ,$$

and

$$P\left(T^+ \mid \overline{L}\right) = 1 - P\left(T^- \mid \overline{L}\right) = 0.01,$$

we obtain

$$P\left(\overline{L} \mid T^+\right) = \frac{(0.01) \cdot (0.999999)}{(0.01) \cdot (0.999999) + (0.98) \cdot (0.000001)} \approx 0.99991.$$

In particular, this implies that

$$P\left(L \mid T^+\right) = 1 - P\left(\overline{L} \mid T^+\right) \approx 0.00009.$$

(b) Try to compute $P\left(L \mid T^-\right)$ in an analogous manner as an exercise.

(Answer: 2.020204×10^{-8}.)